Advances in Wood Processing Technology

Advances in Wood Processing Technology

Martin Kučerka
Alena Očkajová
Richard Kminiak

Basel • Beijing • Wuhan • Barcelona • Belgrade • Novi Sad • Cluj • Manchester

Editors

Martin Kučerka
Department of Technology
Matej Bel University
Banská Bystrica
Slovakia

Alena Očkajová
Department of Technology
Matej Bel University
Banská Bystrica
Slovakia

Richard Kminiak
Department of Woodworking
Technical University in Zvolen
Zvolen
Slovakia

Editorial Office
MDPI AG
Grosspeteranlage 5
4052 Basel, Switzerland

This is a reprint of articles from the Special Issue published online in the open access journal *Applied Sciences* (ISSN 2076-3417) (available at: www.mdpi.com/journal/applsci/special_issues/ Advances_in_Wood_Processing_Technology).

For citation purposes, cite each article independently as indicated on the article page online and as indicated below:

Lastname, A.A.; Lastname, B.B. Article Title. *Journal Name* **Year**, *Volume Number*, Page Range.

ISBN 978-3-7258-2578-3 (Hbk)
ISBN 978-3-7258-2577-6 (PDF)
doi.org/10.3390/books978-3-7258-2577-6

© 2024 by the authors. Articles in this book are Open Access and distributed under the Creative Commons Attribution (CC BY) license. The book as a whole is distributed by MDPI under the terms and conditions of the Creative Commons Attribution-NonCommercial-NoDerivs (CC BY-NC-ND) license.

Contents

About the Editors vii

Preface ix

Martin Kučerka, Alena Očkajová and Richard Kminiak
Special Issue on Advances in Wood Processing Technology
Reprinted from: *Appl. Sci.* **2024**, *14*, 7863, doi:10.3390/app14177863 1

Pavol Koleda, Peter Koleda, Mária Hrčková, Martin Júda and Áron Hortobágyi
Experimental Granulometric Characterization of Wood Particles from CNC Machining of Chipboard
Reprinted from: *Appl. Sci.* **2023**, *13*, 5484, doi:10.3390/app13095484 4

Áron Hortobágyi, Peter Koleda, Pavol Koleda and Richard Kminiak
Effect of Milling Parameters on Amplitude Spectrum of Vibrations during Milling Materials Based on Wood
Reprinted from: *Appl. Sci.* **2023**, *13*, 5061, doi:10.3390/app13085061 19

Marina Chavenetidou and Vasiliki Kamperidou
Impact of Wood Structure Variability on the Surface Roughness of Chestnut Wood
Reprinted from: *Appl. Sci.* **2024**, *14*, 6326, doi:10.3390/app14146326 34

Akbar Rostampour-Haftkhani, Farshid Abdoli, Mohammad Arabi, Vahid Nasir and Maria Rashidi
Effect of Wood Densification and GFRP Reinforcement on the Embedment Strength of Poplar CLT
Reprinted from: *Appl. Sci.* **2023**, *13*, 12249, doi:10.3390/app132212249 48

Abdullah Beram
Enhancing Surface Characteristics and Combustion Behavior of Black Poplar Wood through Varied Impregnation Techniques
Reprinted from: *Appl. Sci.* **2023**, *13*, 11482, doi:10.3390/app132011482 62

Aiuba Suleimana, Bárbara C. Peixoto, Jorge M. Branco and Aires Camões
Experimental Evaluation of Glulam Made from Portuguese Eucalyptus
Reprinted from: *Appl. Sci.* **2023**, *13*, 6866, doi:10.3390/app13126866 76

Zdzisław Kwidziński, Luďka Hanincová, Eryka Tyma, Joanna Bednarz, Łukasz Sankiewicz and Bartłomiej Knitowski et al.
The Efficiency of Edge Banding Module in a Mass Customized Line for Wooden Doors Production
Reprinted from: *Appl. Sci.* **2022**, *12*, 12510, doi:10.3390/app122412510 87

Ivan Ružiak, Rastislav Igaz, Ivan Kubovský, Milada Gajtanska and Andrej Jankech
Prediction of the Effect of CO_2 Laser Cutting Conditions on Spruce Wood Cut Characteristics Using an Artificial Neural Network
Reprinted from: *Appl. Sci.* **2022**, *12*, 11355, doi:10.3390/app122211355 100

Bogdan Warcholinski and Adam Gilewicz
Surface Engineering of Woodworking Tools, a Review
Reprinted from: *Appl. Sci.* **2022**, *12*, 10389, doi:10.3390/app122010389 112

Maciej Sydor, Grzegorz Pinkowski, Martin Kučerka, Richard Kminiak, Petar Antov and Tomasz Rogoziński
Indentation Hardness and Elastic Recovery of Some Hardwood Species
Reprinted from: *Appl. Sci.* **2022**, *12*, 5049, doi:10.3390/app12105049 . . . . **131**

About the Editors

Martin Kučerka

Martin Kučerka is an expert in environmental aspects of wood processing and chip formation analysis. He has worked for many years in the field of wood processing technology and environmental aspects of wood processing. He is the co-author of more than 100 publications with more than 500 citations (including more than 200 in the WoS and Scopus databases). He has been a leader and representative in various woodworking projects. He has extensive expertise and consultation experience with direct applications in economic and social practices. He has provided his expertise and consultancy at a few foreign universities in Zielona Góra, Rzeszow, Ostrava, Brno, Krakow, Prague, and Poznań. He has been cooperating on research with foreign departments at the universities of Poznań, Brno, Sofia, and Prague for a long time.

Alena Očkajová

Alena Očkajová is an expert in the field of environmental aspects of wood processing and chip formation analysis (Department of Technology, Faculty of Natural Sciences, Matej Bel University in Banská Bystrica). She is a member of the commissions for inaugural and habilitation proceedings, for the defense of dissertations, in rigorous, resp. attestation commissions. She is a guarantor of several international scientific conferences.

Richard Kminiak

Richard Kminiak works as an assistant professor at the Department of Woodworking, Faculty of Wood Technology, Technical University of Zvolen. Within his pedagogical activity, he focuses on subjects in many fields, i.e., wood cutting and woodworking, wood cutting tools, machines and equipment for wood processing, CNC technology, CNC machine programming, and CIM systems in the woodworking industry. Within the framework of research activities, he deals with the issues of quality of woodworking products, optimization of technical and technological parameters of woodworking processes, and application of new technologies and technological procedures within woodworking processes.

Preface

This Special Issue, entitled "Advances in Wood Processing Technology", focuses on high-quality original research articles and reviews on the latest (innovative) approaches in the development of wood-based material processing, new ecological wood-based composites, advanced wood processing functions, and further advances in industrial production research in the field of wood-based materials and their applications. Wood is an attractive construction material and has several favorable properties and advantages. However, it also has shortcomings and various limitations, such as limited fire resistance, dimensional instability, and susceptibility to various biotic and abiotic damages. Today, it is possible to improve some properties of wood with the help of extensive innovative methods. Technical progress is largely conditioned by the process of discovering and improving technological processes and methods. New ideas do not usually arise by chance, but social conditions are necessary for this (achieving a certain state of knowledge and technology that requires a new quality, the accumulation of knowledge about a given problem). We are very grateful for contributions concerning the latest technologies for the processing of both growing wood and wood-based composites.

Martin Kučerka, Alena Očkajová, and Richard Kminiak
Editors

Editorial

Special Issue on Advances in Wood Processing Technology

Martin Kučerka [1,*], Alena Očkajová [1] and Richard Kminiak [2]

1 Faculty of Natural Sciences, Matej Bel University, 974 09 Banská Bystrica, Slovakia; alena.ockajova@umb.sk
2 Department of Woodworking, Faculty of Wood Sciences and Technology, Technical University in Zvolen, T. G. Masaryka 24, 960 01 Zvolen, Slovakia; xkminiak@tuzvo.sk
* Correspondence: martin.kucerka@umb.sk

An Overview of Published Articles

The primary goal of this Special Issue, "Advances in Wood Processing Technology", was to showcase cutting-edge research and development in the field of wood-based materials. It aimed to promote innovative approaches for processing wood, creating novel ecological composites, enhancing wood processing functions, and optimizing industrial production processes. Ultimately, this Special Issue sought to contribute to the advancement of wood as a sustainable and high-performance construction material by addressing the challenges and limitations of traditional wood-based materials.

Wood, a naturally abundant and renewable resource, offers numerous advantages as a construction material, including strength, durability, and aesthetic appeal. However, its susceptibility to fire, dimensional instability, and biological degradation poses challenges for its wider application [1]. This Special Issue focused on addressing these limitations through technological innovation [2]. By exploring new processing techniques, developing advanced wood-based composites, and optimizing industrial production, researchers aimed to unlock the full potential of wood as a high-performance material [3].

The first article [4] concludes that wood density directly affects its Brinell hardness and capacity for self-re-deformation. Specifically, denser woods like beech have a lower percentage of permanent indentation, indicating a higher ability to recover shape. This recovery ability in radial and tangential sections is influenced by both density and applied force, whereas in longitudinal sections, it is solely dependent on density. This study highlights that side compression of wood cells is mostly reversible, contrasting with the irreversible damage caused by longitudinal deformation.

The aim of another paper [5] was to investigate the possibilities of increasing the lifetime of woodworking tools through the application of thin hard layers. These layers, applied by physical or chemical vapor deposition methods, have the potential to significantly increase the wear resistance of tools. The results of this study show that multi-layer coatings, especially chromium-based coatings, are the most promising. Coatings with a lower coefficient of friction were found to exhibit higher wear resistance. This property has been shown to be a more important predictor of tool life than the hardness of the coating itself.

The primary goal of the article [6] is to optimize the CO_2 laser cutting process for spruce wood by using an artificial neural network (ANN) to predict cut characteristics. This article specifically focuses on identifying the critical role of laser performance and cutting speed in wood cutting quality and efficiency, developing an ANN model to predict cutting kerf properties and the heat-affected zone based on laser power and the wood's annual ring count. The next section addresses the verification of the model predictions against the existing literature [7–10] and finally proposes the determination of the optimal laser power to achieve the desired quality of spruce wood cut.

The study [11] analyzed the performance of the edge banding module in a fully automated wooden door production line. By examining production data, researchers aimed

Citation: Kučerka, M.; Očkajová, A.; Kminiak, R. Special Issue on Advances in Wood Processing Technology. *Appl. Sci.* **2024**, *14*, 7863. https://doi.org/10.3390/app14177863

Received: 6 August 2024
Revised: 27 August 2024
Accepted: 2 September 2024
Published: 4 September 2024

Copyright: © 2024 by the authors. Licensee MDPI, Basel, Switzerland. This article is an open access article distributed under the terms and conditions of the Creative Commons Attribution (CC BY) license (https://creativecommons.org/licenses/by/4.0/).

to identify factors influencing module efficiency. Results indicate that the module operates flexibly and independently of control parameters, suggesting potential for operational improvements and optimized work scheduling across the entire production line.

The article [12] by Koleda et al. was focused on the determination of the size of wood dust particles generated during milling of chipboard using an experimental optical method. The results showed that the proposed optical method allows more accurate determination of particle size and shape compared to the traditional sieve method. This method provides more detailed information on the composition of wood dust, which can be useful for further research and applications.

The article by Hortobágyi et al. [13] investigated the feasibility of using vibration monitoring on a pneumatic gripper for adaptive control during the milling process. This study analyzed vibration data collected during milling operations, employing fast Fourier transform to identify the dominant frequencies related to tool cutting [14,15]. Statistical analysis revealed that tool type, spindle rotation, and material significantly influenced vibration patterns. While feed rate showed less impact, the results suggest that vibration monitoring can potentially serve as a valuable signal for adaptive control. Future research will focus on combining vibration data with surface roughness measurements and developing a smart pneumatic gripper for the woodworking industry.

The research by Suleiman et al. [16] shows the potential to increase the value of Portuguese eucalyptus forests by using them for glulam production. This research succeeded in producing glulam from eucalyptus without changing the standard production process. The resulting glulam exceeded the strength requirements for high-strength softwood glulam.

The aim of the paper by Abdullah Beram [17] was to improve the properties of black poplar wood through impregnation with calcium oxide solutions. This research compared vacuum and immersion impregnation methods, using different solution concentrations. The results showed that both methods improved the wood's thermal stability and flame resistance while reducing water absorption and swelling. The impregnation process also affected the wood's surface properties, increasing roughness but decreasing water contact angle. This study concludes that calcium oxide is effective in improving the overall performance of black poplar wood.

The study [18] compared the embedment strength of cross-laminated timber (CLT) reinforced with glass fiber-reinforced polymer (GFRP) or densified wood. The results showed that both reinforcement methods improved embedment strength, with densified wood offering the highest values. The embedment strength was higher when the load was applied parallel to the outer layers of CLT [19]. The NDS formula (National Design Specification) provided the most accurate prediction of embedment strength compared to other models. This study concludes that both densification and GFRP reinforcement can enhance CLT performance, but densification appears to be a more effective option.

A recent paper by authors Marina Chavenetidou and Vasiliki Kamperidou [20] investigated the surface roughness of chestnut wood in different anatomical regions and trunk heights. This research found that surface roughness varied significantly depending on the orientation of the wood grain, with measurements perpendicular to the grain resulting in higher roughness values. While no significant differences were observed between heartwood and sapwood roughness, tangential surfaces exhibited the highest roughness overall. This study concludes that surface roughness in chestnut wood is primarily influenced by wood grain orientation and is consistent across different trees and trunk heights.

Funding: This research was supported by the Scientific grant agency of the Ministry of Education, Science, Research and Sports of the Slovak Republic and the Slovak Academy of Sciences under project No. 1/0323/23.

Acknowledgments: Thanks to all the authors and peer reviewers for their valuable contributions to this Special Issue 'Advances in Wood Processing Technology'. I would also like to express my gratitude to all the staff and people involved in this Special Issue.

Conflicts of Interest: The authors declare no conflict of interest.

References

1. Hoadley, R.B. *Understanding Wood: A Craftsman's Guide to Wood Technology*; Taunton Press: Newtown, CT, USA; Publishers Group West [distributor]: Emeryville, CA, USA, 2000; ISBN 978-1-56158-358-4.
2. Kwidziński, Z.; Bednarz, J.; Pędzik, M.; Sankiewicz, Ł.; Szarowski, P.; Knitowski, B.; Rogoziński, T. Innovative Line for Door Production TechnoPORTA—Technological and Economic Aspects of Application of Wood-Based Materials. *Appl. Sci.* **2021**, *11*, 4502. [CrossRef]
3. Braccesi, L.; Monsignori, M.; Nesi, P. Monitoring and Optimizing Industrial Production Processes. In Proceedings of the Proceedings of the Ninth IEEE International Conference on Engineering of Complex Computer Systems, Florence, Italy, 16 April 2004; pp. 213–222.
4. Sydor, M.; Pinkowski, G.; Kučerka, M.; Kminiak, R.; Antov, P.; Rogoziński, T. Indentation Hardness and Elastic Recovery of Some Hardwood Species. *Appl. Sci.* **2022**, *12*, 5049. [CrossRef]
5. Warcholinski, B.; Gilewicz, A. Surface Engineering of Woodworking Tools, a Review. *Appl. Sci.* **2022**, *12*, 10389. [CrossRef]
6. Ružiak, I.; Igaz, R.; Kubovský, I.; Gajtanska, M.; Jankech, A. Prediction of the Effect of CO_2 Laser Cutting Conditions on Spruce Wood Cut Characteristics Using an Artificial Neural Network. *Appl. Sci.* **2022**, *12*, 11355. [CrossRef]
7. Kubovský, I.; Kačík, F.; Reinprecht, L. The Impact of UV Radiation on the Change of Colour and Composition of the Surface of Lime Wood Treated with a CO_2 Laser. *J. Photochem. Photobiol. A Chem.* **2016**, *322–323*, 60–66. [CrossRef]
8. Kúdela, J.; Kubovský, I.; Andrejko, M. Surface Properties of Beech Wood after CO2 Laser Engraving. *Coatings* **2020**, *10*, 77. [CrossRef]
9. Fukuta, S.; Nomura, M.; Ikeda, T.; Yoshizawa, M.; Yamasaki, M.; Sasaki, Y. UV Laser Machining of Wood. *Eur. J. Wood Prod.* **2016**, *74*, 261–267. [CrossRef]
10. Eltawahni, H.A.; Olabi, A.G.; Benyounis, K.Y. Investigating the CO_2 Laser Cutting Parameters of MDF Wood Composite Material. *Opt. Laser Technol.* **2011**, *43*, 648–659. [CrossRef]
11. Kwidziński, Z.; Hanincová, L.; Tyma, E.; Bednarz, J.; Sankiewicz, Ł.; Knitowski, B.; Pędzik, M.; Procházka, J.; Rogoziński, T. The Efficiency of Edge Banding Module in a Mass Customized Line for Wooden Doors Production. *Appl. Sci.* **2022**, *12*, 12510. [CrossRef]
12. Koleda, P.; Koleda, P.; Hrčková, M.; Júda, M.; Hortobágyi, Á. Experimental Granulometric Characterization of Wood Particles from CNC Machining of Chipboard. *Appl. Sci.* **2023**, *13*, 5484. [CrossRef]
13. Hortobágyi, Á.; Koleda, P.; Koleda, P.; Kminiak, R. Effect of Milling Parameters on Amplitude Spectrum of Vibrations during Milling Materials Based on Wood. *Appl. Sci.* **2023**, *13*, 5061. [CrossRef]
14. Asilturk, I. On-Line Surface Roughness Recognition System by Vibration Monitoring in CNC Turning Using Adaptive Neuro-Fuzzy Inference System (ANFIS). *Int. J. Phys. Sci.* **2011**, *6*, 5353–5360. [CrossRef]
15. García Plaza, E.; Núñez López, P.J.; Beamud González, E.M. Efficiency of Vibration Signal Feature Extraction for Surface Finish Monitoring in CNC Machining. *J. Manuf. Process.* **2019**, *44*, 145–157. [CrossRef]
16. Suleimana, A.; Peixoto, B.C.; Branco, J.M.; Camões, A. Experimental Evaluation of Glulam Made from Portuguese Eucalyptus. *Appl. Sci.* **2023**, *13*, 6866. [CrossRef]
17. Beram, A. Enhancing Surface Characteristics and Combustion Behavior of Black Poplar Wood through Varied Impregnation Techniques. *Appl. Sci.* **2023**, *13*, 11482. [CrossRef]
18. Rostampour-Haftkhani, A.; Abdoli, F.; Arabi, M.; Nasir, V.; Rashidi, M. Effect of Wood Densification and GFRP Reinforcement on the Embedment Strength of Poplar CLT. *Appl. Sci.* **2023**, *13*, 12249. [CrossRef]
19. Sydor, M.; Rogoziński, T.; Stuper-Szablewska, K.; Starczewski, K. The Accuracy of Holes Drilled in the Side Surface of Plywood. *BioRes* **2019**, *15*, 117–129. [CrossRef]
20. Chavenetidou, M.; Kamperidou, V. Impact of Wood Structure Variability on the Surface Roughness of Chestnut Wood. *Appl. Sci.* **2024**, *14*, 6326. [CrossRef]

Disclaimer/Publisher's Note: The statements, opinions and data contained in all publications are solely those of the individual author(s) and contributor(s) and not of MDPI and/or the editor(s). MDPI and/or the editor(s) disclaim responsibility for any injury to people or property resulting from any ideas, methods, instructions or products referred to in the content.

Article

Experimental Granulometric Characterization of Wood Particles from CNC Machining of Chipboard

Pavol Koleda [1,*], Peter Koleda [1], Mária Hrčková [1], Martin Júda [2] and Áron Hortobágyi [1]

[1] Faculty of Technology, Technical University in Zvolen, Študentská 26, 960 01 Zvolen, Slovakia; peter.koleda@tuzvo.sk (P.K.); hrckova@tuzvo.sk (M.H.); a.hortobagyi.a@gmail.com (Á.H.)
[2] Faculty of Wood Sciences and Technology, Technical University in Zvolen, T. G. Masaryka 24, 960 01 Zvolen, Slovakia; xjuda@is.tuzvo.sk
* Correspondence: pavol.koleda@tuzvo.sk

Abstract: The aim of this paper is to determine the particle size composition of the wood particles obtained from CNC milling the chipboard using an experimental optical granulometric method. Composite materials (chipboard) are the most-used materials in the woodworking and furniture industries. The proposed optical method of measuring particles' dimensions is compared to the sieving technique. The researched experimental method allows for the determination of not only the size of the fraction of an individual particle's fraction but also more detailed information about the analyzed wood dust emission, for example, the largest and smallest dimension of each single particle; its circularity, area, perimeter, eccentricity, and convex hull major and minor axis length; or the color of the particle.

Keywords: wood particles; optical analysis; MATLAB

Citation: Koleda, P.; Koleda, P.; Hrčková, M.; Júda, M.; Hortobágyi, Á. Experimental Granulometric Characterization of Wood Particles from CNC Machining of Chipboard. *Appl. Sci.* **2023**, *13*, 5484. https://doi.org/10.3390/app13095484

Academic Editor: Stefano Invernizzi

Received: 7 March 2023
Revised: 9 April 2023
Accepted: 26 April 2023
Published: 28 April 2023

Copyright: © 2023 by the authors. Licensee MDPI, Basel, Switzerland. This article is an open access article distributed under the terms and conditions of the Creative Commons Attribution (CC BY) license (https://creativecommons.org/licenses/by/4.0/).

1. Introduction

Machining is one of the highly used processes in modern-day industrial applications, with increasing demands from customers all over the world in areas such as transportation, medicine, surgery, automobiles, space, aeronautics, etc. [1]. The development of industry and the economy puts pressure on the speed of production for products and goods. In addition to the advantages, this trend also has many negative aspects, for example, the excessive production of waste material that can largely be recycled. Waste in the form of particles is also generated during the production of semi-finished products. This is mainly particles and dust, which must be removed so that they do not affect the production process and the operators.

In the specialist literature, sawdust is characterized as a polydisperse bulk material consisting of coarse and medium-coarse fractions, i.e., a bulk material with grain sizes above 0.3 mm, while the proportion of finer fractions with smaller chip sizes is not excluded. According to the classification indicators of bulk materials stated in STN 26 0070, sawdust is classified as B-45UX i.e., a fine-grained loose mass (0.5 ÷ 3.5 mm) that is hygroscopic, and low-flowing and an abrasive mass with a tendency to clump [2].

Sawdust can be used as a secondary raw material. It is one of the starting raw materials for the production of agglomerated chip materials and the chemical processing of wood, a valuable raw material for energy use by direct combustion, and the basic raw material for the production of dimensionally and energetically homogenized fuel (briquettes and pellets) [3].

The carcinogenic risk to humans posed by occupational exposures to wood dust and formaldehyde needs to be evaluated, since a number of occupational situations that involve exposure to wood dust also entail exposure to formaldehyde, such as in plywood and particleboard manufacturing, furniture- and cabinet-making, and parquet floor sanding and varnishing. The highest occupational exposures were noted to occur in wood furniture and cabinet manufacturing, especially during machine sanding and similar operations, in the finishing departments of plywood and particleboard mills, and in the workroom air of sawmills and planer mills near chippers, saws, and planers. Citing findings from several recent well-designed case-control studies, this study concludes that occupational exposure to wood dust is causally related to adenocarcinoma of the nasal cavities and paranasal sinuses [4–6].

Wood processing into a final product is a very complex technological process. The main aim of wood processing is to create a workpiece with the required shape, dimensions, and surface quality [7]. One of the most-used methods of woodworking is milling [8–13]. The quality of the processed surface by milling is affected by various factors, such as the cutting conditions, the blunting of the tool, and the appropriately chosen tool [14–18]. In the case of the wear of the tool during a long period of milling, the vibration frequency may increase, resulting in a decrease in the quality of the milled surface. Tool wear is affected by many factors including the workpiece material, cutting parameters, tool geometry and materials, tool temperature, and cooling methods. All these parameters affect the service life of the tool [19–21].

Currently, the chipboard production is a priority direction in the development of the woodworking industry. Particle board (chipboard) is a material used in the production of cabinet furniture and construction. The popularity of chipboard is also due to the fact that manufacturers of this board material are trying to introduce the promising developments of scientists, to keep up with the times [22,23]. The technology for the production of particle boards is a complex process including a number of important operations. The quality indicators of the finished product largely depend on it. For this reason, we chose this material as the material for our experiment. In addition to the purely technological aspects, the environmental safety of chipboard production is the most relevant, which is reflected in the modern patent, scientific, and technical literature. Chipboard is composed of particles and thin slivers of wood that are made by cutting the wood feedstock with rotating knives and shearing the wood into small elements. The characteristics of chipboard are its low cost, its high thickness, and the capability to manufacture large-dimension boards. Chipboard manufactured from waste materials has an extra carbon offset value, making a contribution to a sustainable environment.

Particleboard is a composite panel product consisting of cellulosic particles of various sizes that are bonded together with a synthetic resin or binder under heat and pressure. Particle geometry, resin levels, board density, and manufacturing processes may be modified to produce products suitable for specific end uses. At the time of manufacture, additives can be incorporated to impart specific performance enhancements including greater dimensional stability, increased fire retardancy, and moisture resistance.

Today's particleboard gives industrial users the consistent quality and design flexibility needed for fast, efficient production lines and quality consumer products. Particleboard panels are manufactured in a variety of dimensions and with a wide range of physical properties that provides maximum design flexibility for specifiers and end users.

2. Materials and Methods

2.1. CNC Machine

The experiments were carried out on a 5-axis CNC machining center SCM Tech Z5 manufactured by the company SCM Group, Rimini, Italy.

The basic technical parameters of the machining center given by the manufacturer are provided in Table 1.

Table 1. Technical and technological parameters of CNC machining center SCM Tech Z5.

Parameter	Range
Userful desktop	$3050 \times 1300 \times 300$ mm
X-axis speed	0–70 m·min^{-1}
Y-axis speed	0–40 m·min^{-1}
Z-axis speed	0–15 m·min^{-1}
Vector rate	0–83 m·min^{-1}
Revolutions	600–24,000 rpm
Power	11 kW
Maximum tool diameter	D = 160 mm
Maximum tool length	L = 180 mm

2.2. Tool Parameters

A diamond shank cutter tool with two rows of cutting diamond blades (Diamond Router Cutter Economic Z2 + 1 − D18 × 26L85S = 20 × 50) was used, manufactured by IGM Tools and Machines (Figure 1). The basic technical and technological parameters given by the manufacturer are provided in Table 2. This tool was chosen for its frequent usage in small woodworking companies due to its high tool lifetime and relatively low cost [24–26]. The cutter was used in previous experiments. The usage time was approximately 120 min.

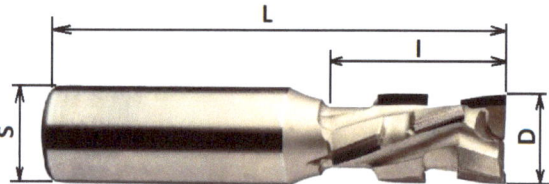

Figure 1. Diamond Shank Cutter Economic Z2 + 1 − D18 × 26L85.

Table 2. Technical parameters of the milling tool.

Name	Working Diameter D (mm)	Working Length L (mm)	Diameter of Chucking Shank S (mm)	Number of Cutting Blades	Material of Cutting Edges
Router Cutter Economic Z2 + 1	18	26	20	2 + 1 HW	Diamond

2.3. Milling Wood Samples

A pressed chipboard was used as a sample for milling. The sample had a raw surface without processing, moisture content of 9.5%, and panel density of 600–640 kg·m^3. Samples of particleboard blanks with the following dimensions, thickness t = 18 mm, width w = 300 mm, and length l = 500 mm, were used in the experiment. The specimens were machined by cylindrical, circumferential milling through the entire thickness, with a diamond shank milling cutter with the following technological parameters: constant depth of cut e = 4 mm; rotation speed of spindle with cutting tool n = 18.000 rpm; feed speed vf = 4, 6, and 8 m·min^{-1}. For each combination of parameters, six specimens in total were collected. The conventional milling (up-milling) method was used for the experiment.

The sawdust obtained during milling was then scanned using a Nikon D5200 camera. This camera was placed on a tripod above the scanned area. The shooting lens was a standard camera lens, Nikon AF-S Nikkor 18–55 mm f/3.5–5.6 GDX VR II (Nikon, Bangkog, Thailand). This lens is designed for use with Nikon's DX-format single-lens reflex cameras. A 3× zoom covers the commonly used focal length range of 18–55 mm and a Silent Wave Motor (SWM) from Nikon offers quiet autofocus. Its view angle is 76–28°50′.

During scanning of the measured sawdust, particles may overlap each other. In this case, the overlapping sawdust would be evaluated as one particle, which would introduce an error into the measurement. So that the sawdust in the sample does not overlap, the particles are separated from each other during the scanning itself using a vibrating table (Figure 2).

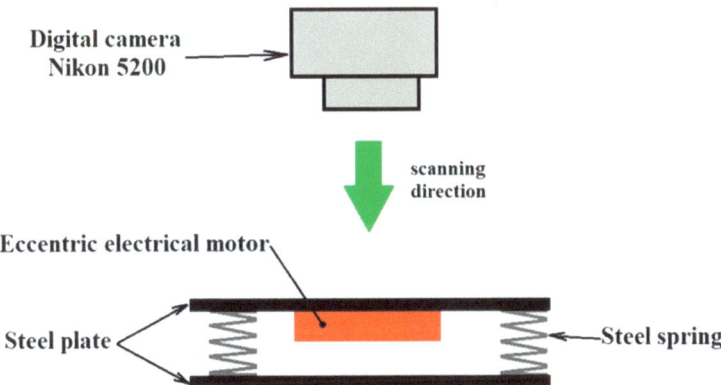

Figure 2. Vibrating table for separating wood particles.

The vibrating table was assembled from two steel plates, which are connected by four springs. The springs were placed in the corners of the plates and fixed by welding. On the bottom of the upper plate, there is an eccentric electric motor in the middle, the movement of which creates an oscillating movement by the upper plate. The speed of the motor and, thus, the strength of the vibrations are adjusted by regulating the supply voltage for the motor. A simple circuit with an LM317 regulator was used as a voltage regulator (Figure 3).

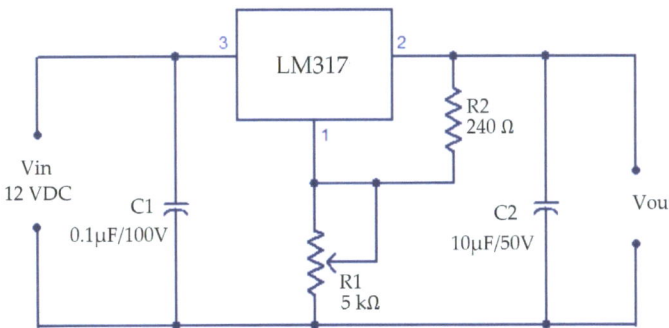

Figure 3. Circuit for voltage regulation. LM317 pins: 1: Adjust; 2: V_{OUT}; 3: V_{IN}.

LM317 is a monolithic integrated circuit in TO-220 packages intended for use as a positive adjustable voltage regulator. It is designed to supply more than 1.5 A of load current, with an output voltage adjustable over a range from 1.2 to 37 V. The nominal output voltage is selected by means of a resistive divider, making the device exceptionally easy to use and eliminating the stocking of many fixed regulators. The input voltage for the controller was 12 VDC voltage from the main adapter. The output voltage from the regulator ranged from 1.25 to 11.3 V. This voltage powered the eccentric motor in the vibrating table.

The particles were scanned with the following parameters (Table 3):

Table 3. Shooting parameters.

Parameter	Value
ISO sensitivity	100
Shutter speed	6.0 s
Aperture	f/5.6
Focal length	55 mm
Effective pixels	24.2 Mpix
Sensor format	APS-C
Image sensor type	CMOS

Scanning of the samples was carried out in low light so that the shadow of the particles was not visible. Therefore, images were recorded with a long exposure, 6 to 15 s.

A sample of an image with analyzed particles is shown in Figure 4.

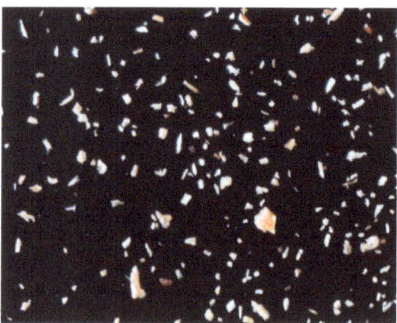

Figure 4. Image with analyzed particles.

Images of sawdust taken in this way were subsequently analyzed in the MATLAB program (MathWorks, Natick, MA, USA), using the proposed program (Figure 5).

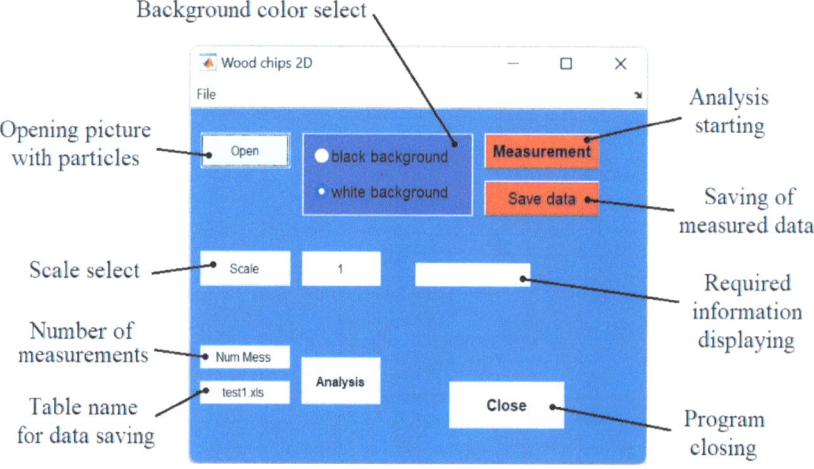

Figure 5. Proposed MATLAB program for wood particles' analysis.

The text block of the displayed required information was mainly used for the design of the application for listing certain information, for example, the value of the content of the test object. During the experiment, the total number of found objects in the currently analyzed image was written into this block.

The "Analysis" button was designed for quick analysis of measured data, such as a histogram of detected particle areas. However, it was not used in the experiment; all analyses were performed in the program Statistica (TIBCO Software Inc., Arlington, VA, USA).

The program works according to the following algorithm (Figure 6).

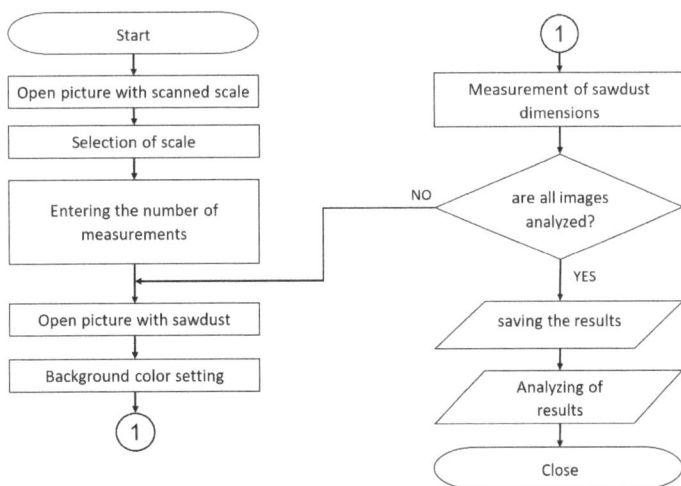

Figure 6. Program algorithm.

First, an image with a reference distance is opened, which helps to determine the scale of the conversion from digital image to the metric system. This step is important to accurately determine the dimensions. The MATLAB program detects all dimensions from the digital image; therefore, the measured values are in pixels. For the conversion to metric dimensions, a conversion coefficient is found. To calculate it, a known distance is scanned in the first opened photo. In the experiment, an office ruler was used. In this ruler, 2 points with a known distance were selected by clicking the mouse. When clicked, the X Y coordinates of the given points were determined, and the distance between the points in pixels was calculated using relationship (1).

$$Dist_{Px} = \sqrt{(X_2 - X_1)^2 + (Y_2 - Y_1)^2} \qquad (1)$$

where

$Dist_{Px}$—distance between the selected points in pixels;
X_1, Y_1—coordinates of the first selected point;
X_2, Y_2—coordinates of the second selected point.

From this distance in pixels in the image and from the known distance on the ruler, the conversion coefficient is then calculated according to relationship (2):

$$Con.Coef = \frac{Dist_{mm}}{Dist_{Px}} \qquad (2)$$

where

Con. Coef—conversion coefficient;
$Dist_{mm}$—known distance in metric system (mm);
$Dist_{Px}$—known distance in pixels.

Subsequently, the analyzed image is opened, in which it is necessary to select the background color of the sawdust, which depends on whether there are light wood particles on a black background or dark wood particles on a white background. These combinations are suitable for the contrast between the searched sawdust and the background, so that they can be easily identified. In the case of a background with a similar color to the searched objects, particles may be incorrectly assigned to the background, or false objects may be created [27,28].

For next analysis, this image is converted into binary form using a function:

$$im2bw(I, graythresh(I)) \tag{3}$$

where *I*—a variable representing the loaded image.

Function im2bw converts the input image to a binary form, in which the pixels belonging to the sawdust have the value of 1 (white), and the other pixels have a value of 0 (black). The decision level for this transfer is calculated using a function: graythresh(I). This computes a global threshold T from grayscale image I, using Otsu's method. Otsu's method chooses a threshold that minimizes the intraclass variance of the thresholded black and white pixels [27–29]. During this binarization, fictitious holes may be created, due to the structure of the sawdust. These are subsequently removed using a function:

$$imfill(BW, 'holes') \tag{4}$$

where

BW—input binary image;
'holes'—parameter of the imfill function.

Function imfill(BW,'holes') fills holes in the input binary image BW. Using parameter 'holes', only holes in objects are removed (Figure 7). Hole is a set of background pixels that cannot be reached by filling in the background from the edge of the image.

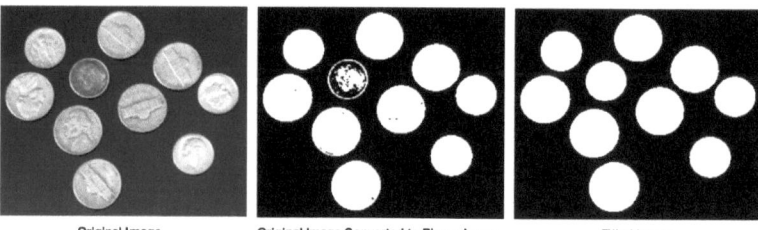

Original Image   Original Image Converted to Binary Image   Filled Image

Figure 7. Removing holes in objects [30,31].

In a binary image modified this way, the dimensional characteristics of the sawdust are subsequently detected by pressing the "Measurement" button.

It is possible to determine the dimensional parameters of the sawdust in the digital image modified in this way. The following functions were used in the MATLAB program to determine sawdust parameters:

$$\begin{array}{c} regionprops(BW, properties) \\ bwferet(BW, properties) \end{array} \tag{5}$$

where

BW—input binary image;
properties—specified, required calculated properties.

Using the regionprops function, the required properties of the found particles are calculated. The list of these characteristics is specified as Properties in the function regionprops. To measure the dimensions of the sawdust, the following were determined: Area, Perimeter, Centroid, Orientation, and Circularity.

Function bwferet measures the Feret properties of objects in an image and returns the measurements in a table. The input properties specify the Feret properties to be measured for each object in input binary image BW. The measured Feret properties include the major and minor axis length, Feret angles, and endpoint coordinates of Feret diameters.

The Feret properties of an object are measured by using boundary points on the antipodal vertices of the convex hull that encloses that object (Figure 8).

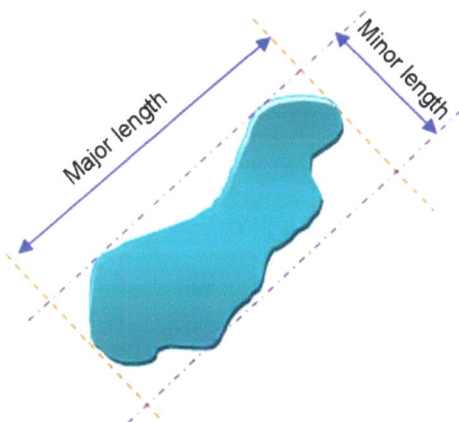

Figure 8. Major and minor length.

Area of individual sawdust is determined with a parameter 'Area'. This parameter counts all pixels belonging to individual sawdust in the binary image.

The Perimeter measurement of sawdust is determined using a parameter 'Perimeter'. Function regionprops computes the perimeter by calculating the distance between each adjoining pair of pixels around the border of the region.

The position of the center of sawdust is determined by a parameter 'Centroid', which detects the horizontal and vertical coordinates of the position of the center of particle in the image.

The rotation of sawdust in the image is detected by a parameter 'Orientation'. This represents an angle between the x-axis and the major axis of the ellipse that has the same second moments as the region, returned as a scalar. The value is in degrees, ranging from $-90°$ to $90°$ [27,28].

Roundness of objects is returned as a structure with parameter 'Circularity'. The structure contains the circularity value for each object in the input image. The circularity value is computed as

$$\frac{4 \cdot Area \cdot \pi}{Perimeter^2} \tag{6}$$

Since MATLAB detects dimensional information about found objects in pixels, the obtained information is converted to metric system. This is accomplished by multiplying the perimeter and the min and max dimension data by the conversion coefficient that was calculated at the beginning of the measurement. Particle area data are multiplied by the square of the coefficient. The circularity parameter is not recalculated by this coefficient because it is a relative quantity. The measured data are sent to an Excel table. The data modified in this way are saved in an Excel table using the "xlswrite" function. The data can be further processed in the Excel program. We exported these data to the program Statistica.

3. Results

The proposed program allows for the measurement of the dimensions of each individual particle. A sample of the determined dimensions is displayed in Figure 9.

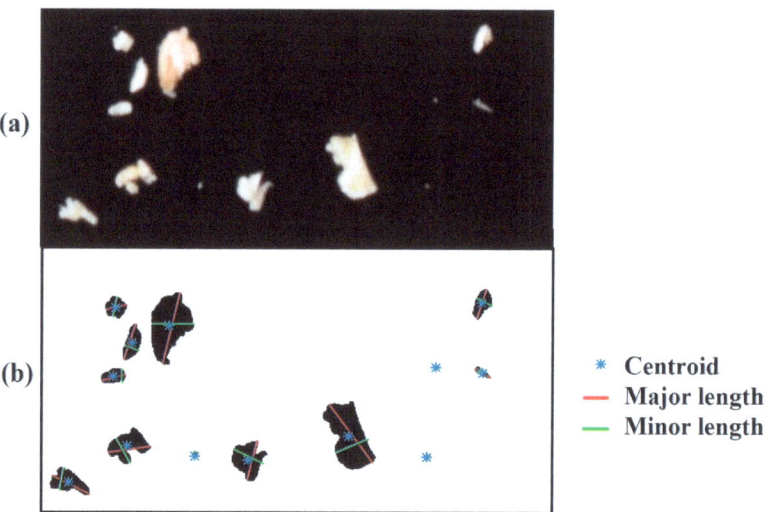

Figure 9. Section from the analyzed image with sawdust: (**a**) original image; (**b**) analyzed image.

The found dimensions were analyzed by one-way ANOVA in the Statistica program (TIBCO Software Inc., USA). Figure 10 shows the weighted means of the area of the analyzed sawdust for individual feed speeds.

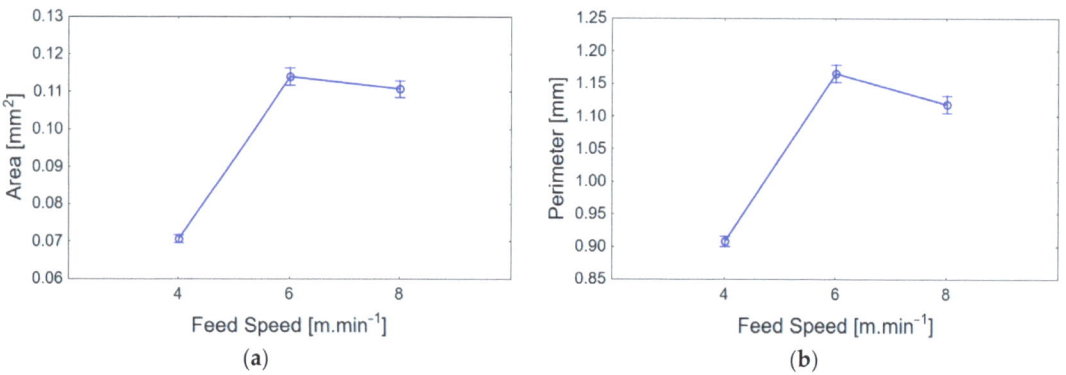

Figure 10. Weighted mean of dimensional characteristics of the analyzed sawdust: (**a**) sawdust area; (**b**) sawdust perimeter.

As shown in Figure 10, the area and perimeter of the sawdust are changing with different feed speeds. The smallest particles were formed by milling with the smallest feed speed. By increasing the feed speed, the size of the generated sawdust also increased. The largest sawdust was created at a feed speed of $v_f = 6$ m·min^{-1}.

The weighted means of the major and minor axes are shown in Figure 11.

To measure by area and perimeter, the major and minor axis dimensions were recorded at a feed speed of 4 m·min^{-1}. At higher feed speeds, the dimensions were larger. The largest sawdust dimensions were recorded at a feed speed of 6 m·min^{-1}.

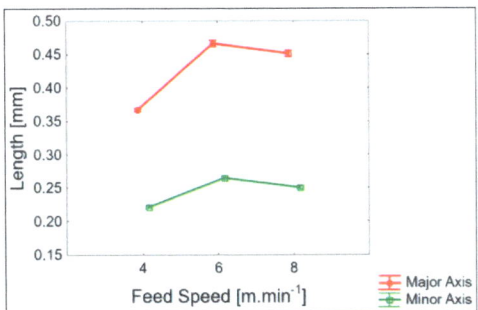

Figure 11. Weighted mean of the largest and smallest dimensions of the analyzed sawdust.

The statistical significance of the detected parameters in the change in speed was determined using Duncan's test (Table 4).

Table 4. Duncan's test of areas.

Feed Speed	{1}	{2}	{3}
4 m·min^{-1}		0.000011	0.000009
6 m·min^{-1}	0.000011		0.011701
8 m·min^{-1}	0.000009	0.011701	

Table 3 shows that the change in feed speed is statistically significant because the probability of the similarity of the datasets is less than 5%.

Using the described method of determining the sawdust dimensions, it is also possible to determine the roundness of objects (Circularity). For a perfect circle, the circularity value is 1.

As shown in Figure 12, the shape of the sawdust is similar to a circle in the sample. For small particles, however, the circularity increases, which is due to the fact that these small particles have a needle-like shape. They have a small area but a larger perimeter. For larger particles, the roundness is smaller because these particles have a shape similar to a circle. The analysis of the variance of the circularity is shown in Figure 13.

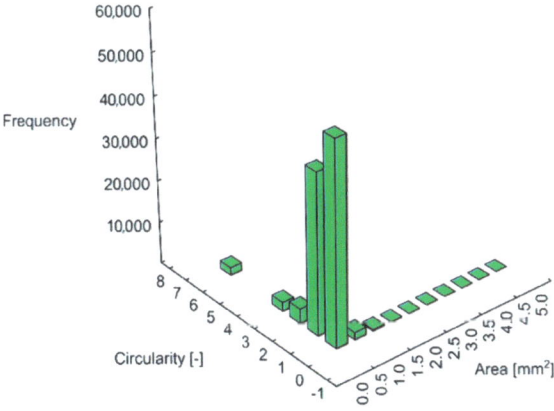

Figure 12. Circularity of sawdust in different areas.

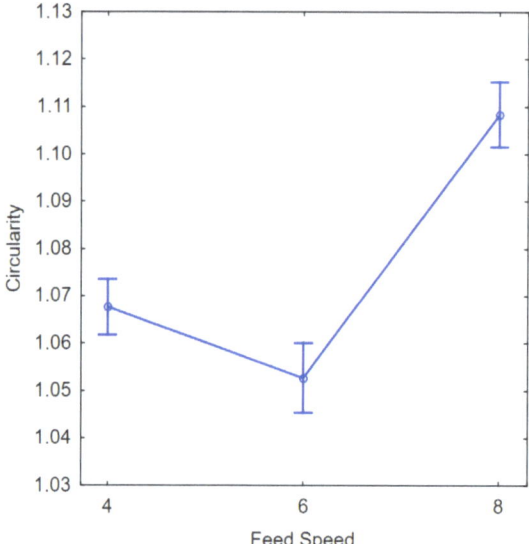

Figure 13. Analysis of variance of circularity.

For a comparison with the sieve analysis, the obtained results were converted from the percentage representation of the sawdust to an area corresponding to the sieves with fractions: 2, 1, 0.5, 0.25, 0.125, 0.063, 0.032, and less than 0.032 mm. Figure 14 shows the results.

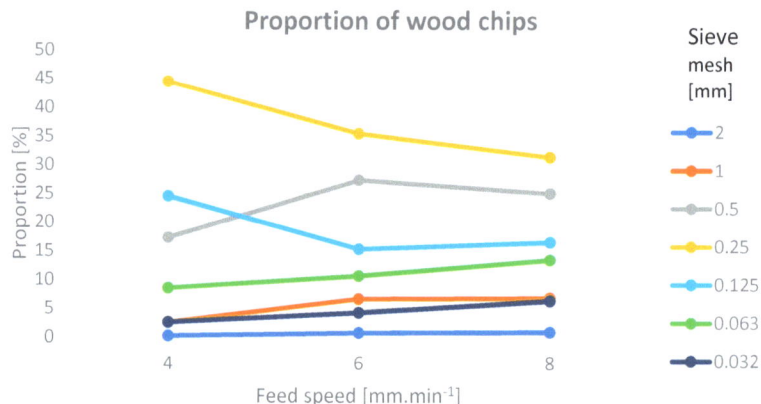

Figure 14. The percentage of individual sawdust fractions.

Figure 14 shows that the largest share of the sawdust was from the fractions 0.25 and 0.5 mm. The major factions (2 a 1 mm) had a small share, approximately 5% at each feed speed.

4. Discussion

An optical analysis of the sawdust and other small materials encounters the problem of overlapping particles during scanning. When several particles overlap each other, such a cluster is typically identified as one separate particle. This problem can be solved using a vibrating table, on which the particles of re ias separated from each other using vibrations before scanning.

Another problem can be capturing all the particles for analysis. The particles' removal efficiency during CNC machining of the particleboard highly depends on the operation type realized by the CNC center as well as the machining pathway [32,33]. For pocketing, the quantity of particles that remained on the panel after routing was negligible and, therefore, the exhaustion efficiency was near 100%. For milling, however, it was at the level of 87%, which was not satisfying. In this study, a particle size analysis of the sawdust was also performed. It showed that the wood particles left over on the machine and around it were not smaller than 0.1 mm. The efficiency of the wood particles' removal decreased with the particles' size increase.

It is possible to identify the effect of shear force on the proportion of the smaller particles within the fine fractions in terms of the influence of the physical and chemical properties of sawed and sanded material, as well as the shape, the dimensions, the sharpness of cutting tools, and technological factors. A feed rate reduction means a decrease in the nominal thickness of the particle, and, thus, the particles move between finer fractions. This fact was also confirmed by other studies [34–36]. The formation of dust particles can occur in all open places of machines as well, especially on the premises of CNC machines as a result of maintenance, repairs, cleaning, inspections, tool changes, etc. [30,31,37–39].

This paper is based on the standard scientific methodologies for the evaluation of particles from the wood milling process, which are accepted for their scientific capacities, but, at the same time, we consider it necessary to discuss this topic from the point of view of objectivity and in the context of the stated findings.

As Kminiak (2021) wrote, it is very difficult to determine the content of the finest dust particles. This content may not be captured by the camera due to the complicated shape of the particles, leading to a possibly incomplete data analysis. Therefore, a smaller focal length camera or a microscope should be used for wood dust with a larger dimensional span. Only then is it possible to detect and quantify the content of the finest dust particles and, thus, to estimate the occupational health risks accordingly [40,41].

One of the ways to improve the chips' geometric measurement is to use the optical method, which was proposed by Sandak et al. (2005) and also by Palubicki et al. (2007). This method has many advantages, since it is simple and fast, does not use very expensive equipment, and has high accuracy [42,43].

5. Conclusions

This research demonstrated the possibility of a more complex analysis of sawdust using the proposed program. In the commonly used methods, for example sieve analyses, the result is only the percentage representation of the size of the individual fractions compared to the total sample [44,45]; thus, the described method allows for obtaining more information about the measured sawdust sample. The dimensional characteristics are determined for each individual particle.

During the analysis of the particles generated during the milling process at different feed speeds, it was found that the smallest particles are generated at a feed speed of 4 m·min^{-1}. The largest sawdust particles were generated at a feed speed of 6 m·min^{-1}. In order to reduce health risks during milling, it is, therefore, not advisable to use low feed speeds. They produce smaller chips and dust that could endanger the health of the operating personnel.

The formation of fine wood dust particles represents a significant occupational hazard to the health and safety of workers. The results obtained can be used for optimizing the technological programs of CNC milling machines, thus reducing the occupational exposure to harmful wood dust emissions in the wood processing industry.

The improvement of the work environment in wood processing and furniture enterprises, by adopting adequate occupational safety and health practices, is desirable not only from the perspective of workers but also because it contributes substantially to labor productivity by enhancing workers' motivation, increasing competitiveness, and promoting economic growth.

During particle scanning, this method was found to be quite time-consuming. This was mainly due to the dark contrasting background of the sawdust. For further experiments, it would be more appropriate to provide light under the sawdust, which would speed up the scanning. Moreover, the particles illuminated in this way would have sharper contours.

In this experiment, a standard lens for a Nikon camera was used. Its maximum focal length is 55 mm. For further research, we plan to try other lenses with longer focal lengths as well as other cameras with extra-long focal lengths. A longer focal length allows for shooting at a smaller view angle. Therefore, the investigated particles scanned in this way should have clearer details.

Author Contributions: Conceptualization, P.K. (Pavol Koleda) and P.K. (Peter Koleda); methodology, P.K. (Pavol Koleda); software, P.K. (Pavol Koleda); validation, P.K. (Pavol Koleda), P.K. (Peter Koleda), and M.H.; formal analysis, P.K. (Pavol Koleda) and M.H.; investigation, P.K. (Pavol Koleda); resources, M.H.; data curation, M.J.; writing—original draft preparation, P.K. (Pavol Koleda); writing—review and editing, P.K. (Pavol Koleda) and Á.H.; visualization, P.K. (Peter Koleda); supervision, P.K. (Pavol Koleda); project administration, P.K. (Pavol Koleda); funding acquisition, P.K. (Peter Koleda). All authors have read and agreed to the published version of the manuscript.

Funding: This research was funded by grant number VEGA 1/0791/21.

Data Availability Statement: Not applicable.

Acknowledgments: This publication is the result of the following projects' implementation: project VEGA 1/0791/21, "Research of non-contact method of analysis of small and dust particles arising in the production process with a prediction of negative effects of dust particles" and project APVV-20-0403, "FMA analysis of potential signals suitable for adaptive control of nesting strategies for milling wood-based agglomerates"; thanks go to the support under the Operational Program Integrated Infrastructure for project II–NITT SK II, "National infrastructure for supporting technology transfer in Slovakia", co-financed by the European Regional Development Fund.

Conflicts of Interest: The authors declare no conflict of interest.

References

1. Danish, M.; Ginta, T.L.; Habib, K.; Carou, D.; Rani, A.M.A.; Saha, B.B. Thermal analysis during turning of AZ31 magnesium alloy under dry and cryogenic conditions. *Int. J. Adv. Manuf. Technol.* **2017**, *91*, 2855–2868. [CrossRef]
2. Hejma, J.; Budinský, K.; Vávra, A. *Vzduchotechnika v Dřevozpracovávajícím Průmyslu*; SNTL: Praha, Czech Republic, 1981; p. 398.
3. Dzurenda, L.; Kučerka, M.; Banski, A. Technologická charakteristika piliny z procesov píleniadreva na rámových, kmeňových pásových a kotúčových pílach. In *Vplyo Techniky na Kvalitu Deleného a Obrábaného Dreva*; Technická Univerzita vo Zvolene: Zvolen, Slovakia, 2008; pp. 93–114. ISBN 978-80-228-1923-7.
4. International Agency for Research on Cancer. *Monographs on the Evaluation of Carcinogenic Risks to Humans*; Wood Dust and Formaldehyde; International Agency for Research on Cancer: Lyon, France, 1995; Volume 62, ISBN 92 832 1262 2.
5. Douwes, J.; Cheung, K.; Prezant, B.; Sharp, M.; Corbine, M.; McLean, D.; 't Mannetje, A.; Schlunssen, V.; Sigsgaard, T.; Kromhout, H.; et al. Wood Dust in Joineries and Furniture Manufacturing: An Exposure Determinant and Intervention Study. *Ann. Work Expo. Health* **2017**, *61*, 416–428. [CrossRef] [PubMed]
6. Nylander, L.A.; Dement, J.M. Carcinogenic effects of wood dust: Review and discussion. *Am. J. Ind. Med.* **1993**, *24*, 619–647. [CrossRef] [PubMed]
7. Gaff, M.; Sarvašová-Kvietková, M.; Gašparík, M.; Slávik, M. Dependence of roughness change and crack formation on parameters of wood surface embossing. *Wood Res.* **2016**, *61*, 163–174.
8. Očkajová, A.; Kučerka, M.; Krišťák, Ľ.; Igaz, R. Granulometric analysis of sanding dust from selected wood species. *Bioresources* **2018**, *13*, 7481–7495. [CrossRef]
9. Vlčková, M.; Gejdoš, M.; Němec, M. Analysis of vibration in wood chipping process. *Akustika* **2017**, *28*, 106–110.
10. Klaric, K.; Greger, K.; Klaric, M.; Andric, T.; Hitka, M.; Kropivsek, J. An Exploratory Assessment of FSC Chain of Custody Certification Benefits in Croatian Wood Industry. *Drv. Ind.* **2016**, *67*, 241–248. [CrossRef]
11. Očkajová, A.; Banski, A. Granularity of sand wood dust from narrow belt sanding machine. *Acta Fac. Xylologiae* **2013**, *55*, 85–90.
12. Kučerka, M.; Očkajová, A. Thermowood and granularity of abrasive wood dust. *Acta Fac. Xylologiae* **2018**, *60*, 43–52.
13. Klement, I.; Vilkovská, T.; Baranski, J.; Konopka, A. The impact of drying and steaming processes on surface color changes of tension and normal beech wood. *Dry. Technol.* **2018**, *37*, 1490–1497. [CrossRef]

14. Welzbacher, C.B.; Brischke, C.; Rapp, A.O. Influence of treatment temperature and duration on selected biological, mechanical, physical and optical properties of thermally modified timber. *Wood Mater. Sci. Eng.* **2008**, *2*, 66–76. [CrossRef]
15. Gejdoš, M.; Lieskovská, M.; Slančík, M.; Němec, M.; Danielová, Z. Storage and fuel quality of coniferous wood chips. *Bioresources* **2015**, *10*, 5544–5553. [CrossRef]
16. Mračková, E.; Krišťák, Ľ.; Kučerka, M.; Gaff, M.; Gajtanska, M. Creation of wood dust during wood processing: Size analysis, dust separation, and occupational health. *Bioresources* **2016**, *11*, 209–222. [CrossRef]
17. Sedlecký, M.; Sarvašová Kvietková, M. Surface waviness of medium-density fibreboard (MDF) and edge-glued panel EGP after edge milling. *Wood Res.* **2017**, *62*, 459–470.
18. Sedlecký, M.; Kvietková, S.M.; Kminiak, R. Medium-density Fiberboard (MDF) and Edge-glued Panels (EGP) after Edge Milling-Surface Roughness after Machining with Different Paraeters. *Bioresources* **2018**, *13*, 2005–2021. [CrossRef]
19. Curti, R.; Marcon, B.; Denaud, L.; Collet, R. Effect of Grain Direction on Cutting Forces and Chip Geometry during Green Beech Wood Machining. *Bioresources* **2018**, *13*, 5491–5503. [CrossRef]
20. Stewart, H.A. High–temperature halogenation of tungsten carbide-cobalt tool material when machining MDF. *For. Prod. J.* **1992**, *42*, 27–31.
21. Kvasnová, P.; Novák, D.; Novák, V. Laser welding of aluminium alloys. *Manuf. Technol.* **2017**, *17*, 892–898. [CrossRef]
22. Nishimura, T. Chipboard, oriented strand board (OSB) and structural composite lumber. In *Wood Composites*; Woodhead Publishing: Sawston, UK, 2015; pp. 103–121. ISBN 9781782424543. [CrossRef]
23. Ischenko, T.L.; Efimova, T.V. Analysis of heat and mass transfer process while cooling particle boards produced on low-toxic resins. *J. Phys. Conf. Ser.* **2022**, *2388*, 012091. [CrossRef]
24. Kučerka, M.; Očkajová, A.; Kminiak, R.; Pędzik, M.; Rogozinski, T. The Effect of the Granulometric Composition of Beech Chips from a CNC Machining Center on the Environmental Separation Technique. *Acta Fac. Xylologiae* **2022**, *64*, 87–97. [CrossRef]
25. Wei, W.H.; Xu, J.H.; Fu, Y.C.; Yang, S.B. Tool wear in turning of titanium alloy after thermohydrogen treatment. *Chin. J. Mech. Eng.* **2012**, *25*, 776–780. [CrossRef]
26. Júda, M.; Šustek, J.; Krišťák, Ľ. Granulometric characteristics of MDF and particleboard by varying technological parameter feed speed using a CNC milling router. *Chip Chipless Woodwork. Process.* **2022**, *13*, 39–49.
27. Koleda, P.; Hrčková, M. Sawdust analysis using Matlab. *Chip Chipless Woodwork. Process.* **2022**, *13*, 59–64.
28. Koleda, P.; Koleda, P. Possibilities of determining the characteristics of woodchips using image analysis in the matlab program. In Proceedings of the Eleventh International Scientific and Technical Conference Innovations in Forest Industry and Engineering Design INNO 2022, Hotle Rila, Borovets, Bulgaria, 3–5 October 2022; Faculty of Forest Industry, University of Forestry: Sofia, Bulgaria, 2022; pp. 133–140, ISBN 978-619-7703-01-6.
29. Otsu, N. A Threshold Selection Method from Gray-Level Histograms. *IEEE Trans. Syst. Man Cybern.* **1979**, *9*, 62–66. [CrossRef]
30. Goli, G.; Fioravanti, M.; Marchal, R.; Uzielli, L. Up-milling and down-milling wood with different grain orientations—Theoretical background and general appearance of the chips. *Eur. J. Wood Prod.* **2009**, *67*, 257–263. [CrossRef]
31. Kminiak, R.; Němec, M.; Izag, R.; Gejdoš, M. Advisability-Selected Parameters of Woodworking with a CNC Machine as a Tool for Adaptive Control of the Cutting Process. *Forests* **2023**, *14*, 173. [CrossRef]
32. Pałubicki, B.; Rogozinski, T. Efficiency of chips removal during CNC machining of particleboard. *Wood Res.* **2016**, *61*, 811–818.
33. Očkajová, A.; Kučerka, M.; Kminiak, R.; Krišťák, Ľ.; Igaz, R.; Réh, R. Occupational Exposure to Dust produced when Milling Thermally ModifiedWood. *Int. J. Environ. Res. Public Health* **2020**, *17*, 1478. [CrossRef]
34. Kminiak, R.; Dzurenda, L. Impact of Sycamore Maple Thermal Treatment on a Granulometric Composition of Chips Obtained due to Processing on a CNC Machining Centre. *Sustainability* **2019**, *11*, 718. [CrossRef]
35. Dzurenda, L.; Wasielewski, R.; Orlowski, K. Granulometric analysis of dry sawdust from sawing process on the frame sawing Machine PRW-15M. *Acta Fac. Xylologiae* **2006**, *48*, 51–57.
36. Kminiak, R.; Banski, A. Granulometric analysis of chips from beech, oak and spruce woodturning blanks produced in the milling process using axial CNC machining center. *Acta Fac. Xylologiae* **2019**, *61*, 75–82.
37. Tureková, I.; Mračková, E.; Marková, I. Determination of Waste Industrial Dust Safety Characteristics. *Int. J. Environ. Res. Public Health* **2019**, *16*, 2103. [CrossRef] [PubMed]
38. Igaz, R.; Kminiak, R.; Krišťák, L.; Němec, M.; Gergeľ, T. Methodology of Temperature Monitoring in the Process of CNC Machining of Solid Wood. *Sustainability* **2019**, *11*, 95. [CrossRef]
39. Kopecký, Z.; Rousek, M. Dustiness in high-speed milling. *Wood Res.* **2007**, *52*, 65–76.
40. Kminiak, R.; Kučerka, M.; Kristak, L.; Reh, R.; Antov, P.; Očkajová, A.; Rogoziński, T.; Pędzik, M. Granulometric Characterization of Wood Dust Emission from CNC Machining of Natural Wood and Medium Density Fibreboard. *Forests* **2021**, *12*, 1039. [CrossRef]
41. Pałubicki, B.; Hlásková, L.; Frömel-Frybort, S.; Rogozinski, T. Feed Force and Sawdust Geometry in Particleboard Sawing. *Materials* **2021**, *14*, 945. [CrossRef]
42. Harrell, M.; Selvaraj, S.A.; Edgar, M. DANGER! Crisis Health Workers at Risk. *Int. J. Environ. Res. Public Health* **2020**, *17*, 5270. [CrossRef]
43. Effect of Selected Additives to Properties of Wood Pellets and Their Production. Available online: https://www.researchgate.net/publication/260225831_Effect_of_selected_additives_to_properties_of_wood_pellets_and_their_production (accessed on 5 April 2023).

44. Pałubicki, B.; Kowaluk, G.; Frąckowiak. Convexity–Additional parameter to chips geometry characterization. *Drewo Wood* **2007**, *51*, 178.
45. Sandak, J.; Orlowski, K.; Negri, M. Chip geometry while sawing frozen wood. In *COST Action E35 Workshop On Procesing Of Frozen Wood*; Lappeenranta University of Technology: Lappeenranta, Finland, 2005.

Disclaimer/Publisher's Note: The statements, opinions and data contained in all publications are solely those of the individual author(s) and contributor(s) and not of MDPI and/or the editor(s). MDPI and/or the editor(s) disclaim responsibility for any injury to people or property resulting from any ideas, methods, instructions or products referred to in the content.

Article

Effect of Milling Parameters on Amplitude Spectrum of Vibrations during Milling Materials Based on Wood

Áron Hortobágyi [1,*], Peter Koleda [1], Pavol Koleda [1] and Richard Kminiak [2]

1. Department of Manufacturing and Automation Technology, Faculty of Technology, Technical University in Zvolen, T. G. Masaryka 24, 960 01 Zvolen, Slovakia
2. Department of Woodworking, Faculty of Wood Sciences and Technology, Technical University in Zvolen, T. G. Masaryka 24, 960 01 Zvolen, Slovakia
* Correspondence: xhortobagyi@is.tuzvo.sk

Abstract: Milling with use of CNC machines is a well-established method and much research was concluded on this topic. However, when it comes to wood and wood composites, the material non-homogeneity brings a lot of variability into cutting conditions. As a part of research into potential signals for nesting milling, material vibrations at clamping points were examined in this study. The main goal was to conclude if cutting parameters have a statistically significant effect on measurement. The place of measurement was analyzed so it was accessible to the machine operator. Medium density fiberboard and particleboard specimens were cut through by razor and spiral mill, with spindle rotating 10,000 and 20,000 min^{-1} and feed rates 2, 6, 10 $m \cdot min^{-1}$. Vibrations were measured at vacuum grippers, and were then processed by fast Fourier transform. Then, frequency spectrum maxima were compared, as well as amplitude sizes. Main frequencies were of roughly 166 Hz and multiples, suggesting their origin in tool rotation. When maxima were compared, tool use, spindle rotation, and feed rate seemed to affect the result. Frequency spectrum amplitudes were subjected to analysis of variance, significant effect was found on spindle speed, tool, and specimen material. No significant effect was found with differing feed rates.

Keywords: MDF; CNC; milling; vibration measurement

1. Introduction

The technology of machining native wood and wood-based materials by multi-axis CNC machining centers is increasingly used especially in the manufacturing of complex parts, or so-called nesting milling. CNC machines are often used without direct control by a human operator and, therefore, the setting of appropriate technological parameters is extremely important for trouble-free machining, achieving the required quality of the workpiece, minimizing vibration and electricity consumption, and, last but not least, ensuring adequate work environment [1,2]. The ongoing digital revolution (Industry 4.0) brings techniques for inspection and for collecting data on the production process and their online processing remotely using interconnected cyber-physical systems. It does not have to be only about evaluating data in real time, but also about their prediction with the aim of environmental protection and sustainability industry [3]. The problems related to this can result in the development of new intelligent sensors, methods, and procedures for measuring quantities that indirectly characterize the machining process (energy consumption, acoustic emissions, vibrations, dust) and are suitable for creating a digital twin of intelligent production [4].

Milling is a well-known manufacturing process, where many problems already have been addressed. Many studies were conducted to find dependence of roughness of machined material on cutting parameters. Generally, the surface roughness depends on spindle speed, feed rate, and tool diameter. To achieve a smoother surface, high spindle

speed with slow federate should be used, ideally with a small diameter tool [5–13]. However, such parameters also create higher cutting forces which are undesirable [14]. A big aspect is from material itself. Especially where wood is concerned, material homogeneity plays a big role.

Medium density fiberboard (MDF) is a wood-based industrial product. It is made from wood waste fibers bonded together by resin, while heated under pressure. MDF has certain advantages when compared to native wood and is currently preferred in many applications [10]. MDF roughness from manufacture increases with rising compression strength [15]. It is generally denser than particle board as well as plywood. Even though it consists of fibers and not veneers, it can be applied as a construction material in most cases, where plywood is used nowadays.

The density of a typical MDF is between 500 kg·m^{-3} and 1000 kg·m^{-3} whereas the density of a particle board is in the range from 160 kg·m^{-3} to 450 kg·m^{-3}. In contrast to natural wood, MDF does not contain knots or rings [16]. Experimental studies provided assessment of the concrete cutting parameters for machining isotropic and orthotropic wood-based materials [17,18].

Vibrations during the machining process were subjected to multiple studies. Vibrations are often frequent problem which affects dimensional accuracy of the parts being machined, surface finish quality and tool life. Vibrations are induced due to machine faults, cutting tool, cutting parameters, workpiece deformation, etc. These vibrations are generally measured using accelerometers mounted on various machine parts elements [19].

From the point of view of vibration measurement, several articles were published, especially in the area of metal materials machining. A relation was found for steel [20] and titanium [21] machining, that surface roughness is most dependent on feed rate, while cutting speed has the highest effect on tool vibration. Similar results were found for aluminum in [22]. Other studies were focused on finding a relation between vibrations and surface finish, with signal spectrum analysis and wavelet packet transform (WPT), where it resulted in vibration ranges correlating the vibration amplitudes with resultant surface roughness. The measured vibration and wavelet packet transform method could be effectively applied for real-time, highly accurate, and reliable roughness monitoring, with a low computation power cost in CNC machining [23]. Singular spectrum analysis was also considered as a viable strategy for assessing vibration signals for the real-time monitoring of machined surface finish [24], as well as the calculation of surface quality in CNC turning by model-assisted response surface approach [25]. Tool vibration signals were experimentally monitored by spectral kurtosis and ICEEMDAN energy modes for insert abrasion assessment [26]. Surface quality prediction models based on a regression method and artificial neural network were developed in [27,28]. Multiple methods involving machine learning were summarized and compared in [29]. Tool geometry was also found to be a big factor, when vibrations are assessed [30–32].

The fast Fourier transform used in this study is a valuable tool in situations when signal processing is needed. It is a computation instrument for easier signal analysis. FFT can be used on computers for power spectrum analysis as well as for filter simulation. This tool is basically an efficient way to calculate the discrete Fourier transform of data sample sequences [33] and to transform the time domain signal into frequency domain [34]. The Fourier transform was applied for surface roughness prediction to obtain the features of image texture [35], and for real-time measurement and intervention system for build-up-edge and tool damage to analyze the vibration signals for fast recognition of signal irregularities [36]. It was also successfully tested for on-the-fly CNC interpolation method [37]. In this research, fast Fourier transform was used to find dominant frequencies which were combined with noise in a composite signal. When multiple frequencies are combined, it is almost impossible to decompose a signal into original elements by signal shape assessment. After fast Fourier transform, two-sided amplitude spectrum and single-sided amplitude spectrum can be calculated with use of signal length. Other examples of the fast Fourier transform use are the conversion of Gaussian pulse or the turning of periodical waves from the time domain to the frequency domain [38].

Experiments with MDF routing showed that a suitable cutting edge angle, with consideration of tool material, is in lower scope of ordinarily used HM tools. From the range of the research, the best performing angle was roughly 40 degrees [39]. Another factor that was both cause and effect of vibration is tool wear [2]. To mitigate this effect, many methods were tested, with a range of them focusing on MDF routing. An adaptive regulation system generating responses to advancing tool wear was developed in [40]. A neural network was tested for tool wear monitoring, where several machining factors were measured, including cutting forces, temperature, and power [41]. Mathematical models were also used for the correction of size parameters at 13 levels of a tool wear, with resulting control charts [42].

Another factor affecting vibration is clamping. Vacuum clamping systems, such as the one used in this study, are usually used for particle boards milling on woodworking machining centers. The vacuum gripper systems provide good access to the workpiece edges during machining. A downside of this securing method is un-clamped board areas with relative distance to nearest gripper. These are relatively free to vibrate in a wide frequency range while machining takes place. Due to these vibrations, the roughness of the machined edge is higher, and the process is accompanied by high acoustic emission [43]. However, all the above could be measured by other non-intrusive methods, such as vibro-acoustic analysis [44–46], or by means of energy consumption tests [47,48].

This research focused on the evaluation of vibrations during milling of medium density fiberboards on 5-axis CNC machine. The goal of the study was a confirmation of the effect of changing the technological parameters of milling wood-based agglomerates on the size of the vibration amplitude. Vibration measurement should serve as one of the appropriate signals for an adaptive machining control system, which is the goal of ongoing research of FMA analysis of potential signals suitable for adaptive control of nesting strategies for milling wood-based agglomerates. The research was intended to confirm the hypotheses that a change in spindle speed, feed rate, workpiece material, and a change in the tool influences the change in the amplitude of the resulting vibrations. At the same time, one of the goals of the article was to measure these vibrations in a place accessible to the operator.

2. Materials and Methods

Medium density fiberboards cut to 500 × 300 × 18 mm with weight 1960 g were used as specimens. The density of these specimens was 720–740 kg·m^{-3}. Particleboards of the same dimensions were added as reference specimens. The density of particleboards given by the manufacturer was 600–640 kg·m^{-3} (deciduous 10%, coniferous 90%), and urea formaldehyde glue with paraffin admixture was used; both originated from Kronospan Ltd., Zvolen, Slovakia. The manufacturer declared that the material complied with the EN 14,322 standard, EN 312-2, and emission class E1 (EN ISO 12460-5) [49–51].

A measurement was conducted on a 5-axis CNC machining center SCM Tech Z5 (Figure 1), in laboratories of Technical University in Zvolen. Table 1 provides the basic technical-technological parameters given by the manufacturer.

Figure 1. CNC machining center SCM Tech Z5.

Table 1. CNC machining center SCM Tech Z5 technical parameters.

CNC Machining Center SCM Tech Z5 Technical Parameters	
Useful desktop (mm)	X = 3050, Y = 1300, Z = 3000
Speed in x axis (m·min^{-1})	0 ÷ 70
Speed in y axis (m·min^{-1})	0 ÷ 40
Speed in z axis (m·min^{-1})	0 ÷ 15
Vector rate (m·min^{-1})	0 ÷ 83
Technical Parameters of the Electric Spindle with HSK F63 Connection	
Rotation in C axis	640°
Rotation in B axis	320°
Revolutions (min^{-1})	600 ÷ 24000
Electric power (kW)	11
Maximum tool dimensions (mm)	D = 160, L = 180

Vibration was measured by PicoScope with MEMS accelerometer TA143 [52] with parameters in Table 2.

Table 2. Basic parameters of accelerometer TA143.

Parameter	Value
Maximum measurable acceleration	±5 g
Output	0–2 V DC
Output scaling	99 to 122 mV·g^{-1}
0 g output	0.85 to 1.15 V

Specimens were clamped to four pneumatic grippers, each with surface 120 × 120 mm and clamping force 16 kg/m^2 (Figure 2). Each cut had a depth of 19 mm, and so, the whole thickness of the material was machined in one run.

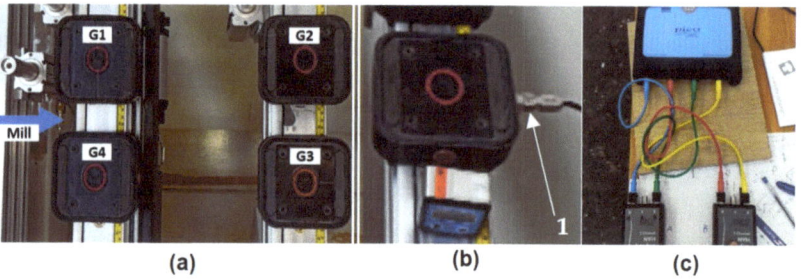

Figure 2. Measurement assembly. (**a**) Pneumatic grippers (G1–G4), (**b**) magnetic attachment of accelerometer (1), (**c**) NVH kit with PicoScope.

To emulate the conditions of nesting milling, a tool trajectory cut through the middle of specimen. The spiral and razor cutter shown in Figure 3 were used as tools, with parameters summed in Table 3.

Varying feed rate of 2, 6, 10 m·min^{-1} was used, with spindle rotations 10,000 and 20,000 min^{-1}. Cutting was repeated 3 times for each parameter. Both experimental layout and cutting parameters are shown in Figure 4 and Table 4 [53].

As seen in Figures 2 and 4, accelerometer probes were attached onto pneumatic grippers. To select placement location, multiple tests were conducted with probes placed on each gripper. An example of a result with spindle revolutions of 20,000 min^{-1} and feed rate of 6 m·min^{-1} is shown in Figure 5.

Although cutting parameters were the same, the placement of accelerometer had a visible impact. Grippers G1 and G4 had a larger peak in the beginning of cutting, as they were closer to the start of tool path. However, there was still a big difference, due to tool rotation. On gripper G1 and G2, vibrations were caused by conventional milling, while grippers G3 and G4 side were experiencing climb milling. Another useful fact was that vibrations in y axis never showed as maximum in measurement.

Therefore, it was concluded that four channels of accelerometer were used to simultaneously measure x and z axes of two grippers. Goal was to have at least one sensor as close to the source as possible, and so, a pair of grippers, one near start and one near end position, were chosen. Lastly, as grippers G3 and G4 showed larger extremes, they were the final choice and sensors were placed as shown in Figure 4.

Data were then processed in MATLAB (MathWorks, Inc., Sentic, MA, USA). The initial signal was cut, so only the milling part would be assessed, and measured voltage was converted to acceleration by output scaling [52].

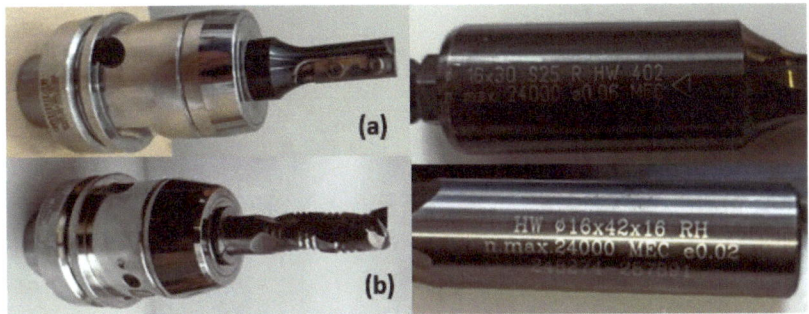

Figure 3. Cutters used in experiment. (**a**) Razor cutter, (**b**) spiral cutter.

Table 3. Tool parameters.

Parameter	Razor Mill	Spiral Mill
Flute diameter	16 mm	16 mm
Shaft diameter	25 mm	16 mm
n max	24,000	24,000
Teeth	2	3
Cut direction	straight	up-cut
Cutting edge	IGM D16 L28.3	solid
Chip breaker	-	yes
Tool carrier	HSK 63 GM 300	HSK 63 GM 300
Reduction sleeve	-	16–25

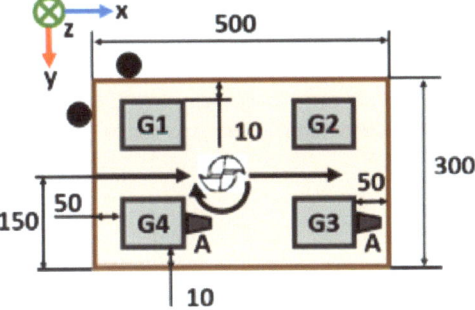

Figure 4. Experiment layout: Specimen placement and tool path on CNC. G—pneumatic gripper, A—accelerometer, x, y, z-orientation of accelerometer axes, G1–G4—Grippers no. 1–4.

Table 4. Number of measurements for each combination of cutting parameters.

Cutting Parameters			VF [m·min⁻¹]								
			2			6			10		
Tool	n [min⁻¹]	Grip.	x	y	z	x	y	z	x	y	z
spiral mill	10,000	1	-	-	-	-	-	-	-	-	-
		2	-	-	-	-	-	-	-	-	-
		3	3	-	3	3	-	3	3	-	3
		4	3	-	3	3	-	3	3	-	3
	20,000	1	-	-	-	3	3	3	-	-	-
		2	-	-	-	3	3	3	-	-	-
		3	3	-	3	3	3	3	3	-	3
		4	3	-	3	3	3	3	3	-	3
razor cutter	10,000	1	-	-	-	-	-	-	-	-	-
		2	-	-	-	-	-	-	-	-	-
		3	3	-	3	3	-	3	3	-	3
		4	3	-	3	3	-	3	3	-	3
	20,000	1	-	-	-	-	-	-	-	-	-
		2	-	-	-	-	-	-	-	-	-
		3	3	-	3	3	-	3	3	-	3
		4	3	-	3	3	-	3	3	-	3

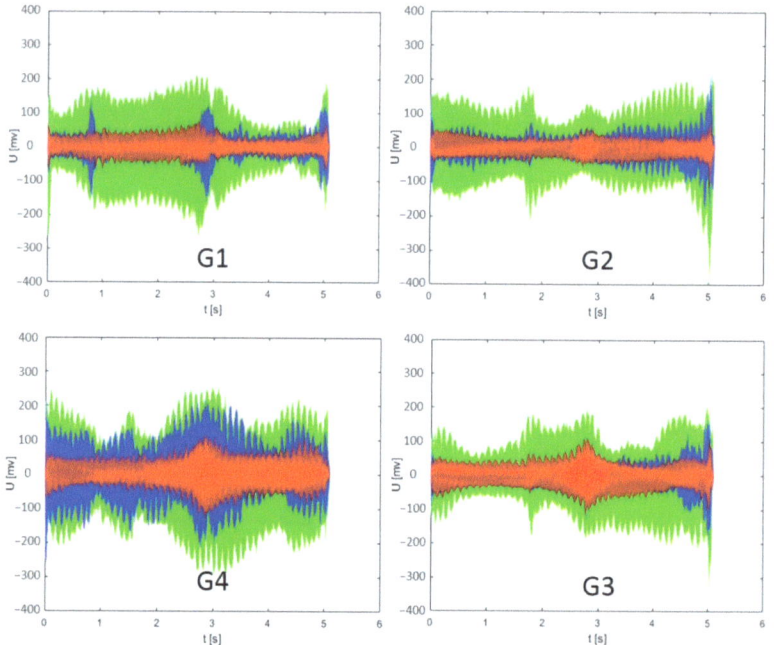

Figure 5. Result of measurement on individual axes. (**G1–G4**)—pneumatic grippers no. 1 to 4, blue—x axis, red—y axis, green—z axis.

Parts of the signal which contained information from before and after the milling process were cut. The threshold value ± 7 mV was used to determine the beginning and ending of the desired signal. Then, fast Fourier transform in MATLAB was used to calculate single-sided amplitude spectrum of vibrations [38] (percent sign "%" marks comments in MATLAB code):

L = length (X);
Fs = sampling frequency;
T = 1/Fs;
T = (0:L−1)*T;
Y = fft(X); % computing of FFT;
P2 = abs(Y/L);
P1 = P2(1:L/2 + 1);
P1(2:end−1) = 2*P1(2:end−1);
f = Fs*(0:(L/2))/L;
plot(f,P1); % generating of FFT graph;
title ("Single-Sided Amplitude Spectrum of S(t)"); % title of generated graph;
xlabel ("f (Hz)"); % x and y labels of generated graph;
ylabel(" | P1(f) | ");

In this code, X is input variable as sequence of acceleration according to the sampling period; L is number of samples in X; Fs is the value of sampling frequency; T is sampling period; t is time of discrete sample according to sampling period; Y is result of FFT from signal X; P2 are absolute values of Y/L ratio; P1 is computing of one-side spectrum amplitudes.

As there were multiple low peaks (noise) in mixture with significantly higher amplitude peaks, a threshold value of $|Y(f)| = 0.5$ was set as minimum for result to be recorded. Additionally, multiple frequency maxima were concentrated around certain values. For better readability, only the maximal value was recorded from such groups. Original signal, its cut form, and dominant frequencies with their maxima are shown in Figure 6.

The entire procedure of the experiment described in previous paragraphs is graphically displayed in Figure 7.

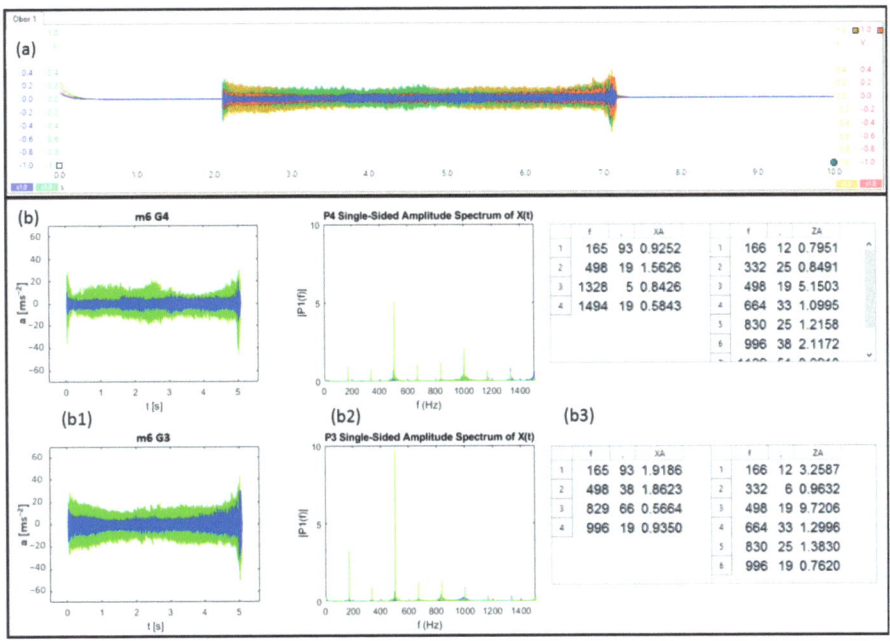

Figure 6. Example of signal processing. (**a**) PicoScope output showing vibrations of four channels, blue—x axis of gripper 4, green—y axis of gripper 4, red—x axis of gripper 3, yellow—y axis of gripper 3, (**b**) MATLAB output, (**b1**) signal from PicoScope (**b2**) Amplitude spectrum as FFT result (**b3**) list of peaks and their maxima, blue—x axis result, green—z axis result.

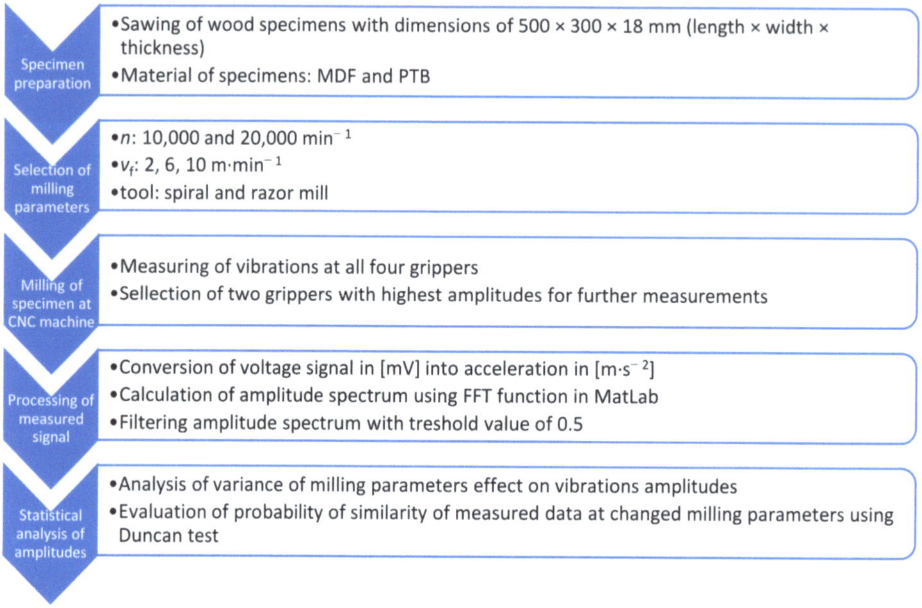

Figure 7. Scheme of experiment procedure.

Specimen preparation
- Sawing of wood specimens with dimensions of 500 × 300 × 18 mm (length × width × thickness)
- Material of specimens: MDF and PTB

Selection of milling parameters
- n: 10,000 and 20,000 min^{-1}
- v_f: 2, 6, 10 m·min^{-1}
- tool: spiral and razor mill

Milling of specimen at CNC machine
- Measuring of vibrations at all four grippers
- Sellection of two grippers with highest amplitudes for further measurements

Processing of measured signal
- Conversion of voltage signal in [mV] into acceleration in [m·s^{-2}]
- Calculation of amplitude spectrum using FFT function in MatLab
- Filtering amplitude spectrum with treshold value of 0.5

Statistical analysis of amplitudes
- Analysis of variance of milling parameters effect on vibrations amplitudes
- Evaluation of probability of similarity of measured data at changed milling parameters using Duncan test

3. Results

To enable data assessment, the results shown in Figure 6(b3) were also programmatically saved to tables, as shown in Table 5. Each dominant frequency is marked by bold font.

Table 5. Example of filtered output from fast Fourier transform.

G3 X		G3 Z		G4 X		G4 Z	
f (Hz)	Maxima	f (Hz)	Maxima	f (Hz)	Maxima	f (Hz)	Maxima
165.93	0.93	166.13	0.80	**165.93**	**1.92**	166.13	3.26
498.20	**1.56**	332.26	0.85	498.39	1.86	332.06	0.96
1328.06	0.84	**498.20**	**5.15**	829.67	0.57	**498.20**	**9.72**
1494.19	0.58	664.33	1.10	996.19	0.94	664.33	1.30
		830.26	1.22			830.26	1.38
		996.39	2.12			996.19	0.76
		1162.52	0.69				
		1328.65	0.51				

Dominant frequencies were summarized. In most cases, when no parameters changed, they remained similar. A summary of their averages is shown in Figure 8.

Next, to determine if vibrations on the gripper could be used as a signal for adaptive control, the dependency between cutting parameters and single-sided amplitude spectrum amplitudes were assessed by analysis of variance (ANOVA). This analysis was conducted for significance level p = 5%. Figure 9 shows the dependency of amplitudes when changing spindle rotation from 10,000 to 20,000 min^{-1}. As the revolutions increased, the vibration amplitudes also increased.

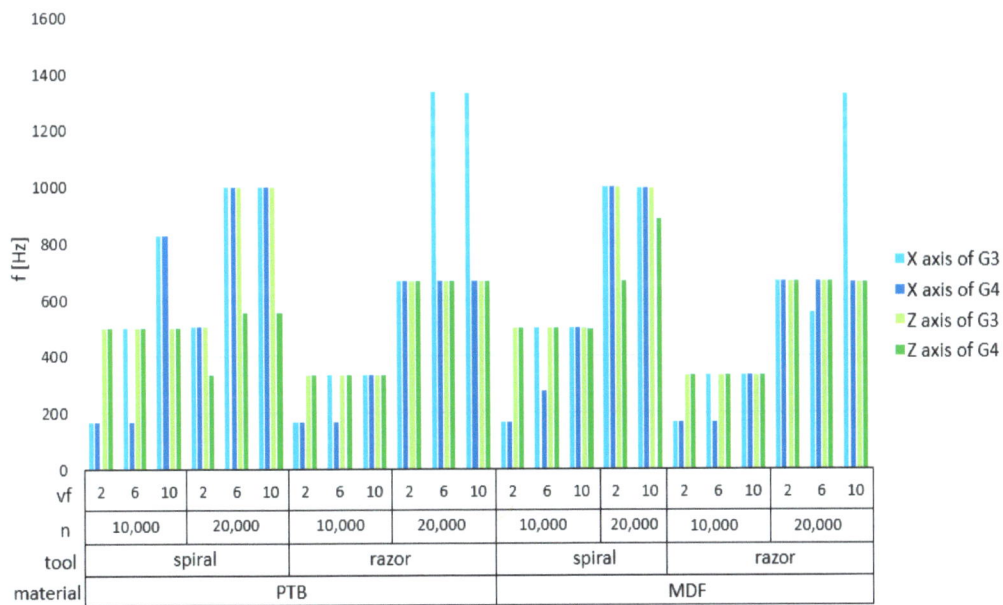

Figure 8. Summary of dominant frequency maxima though range of parameters.

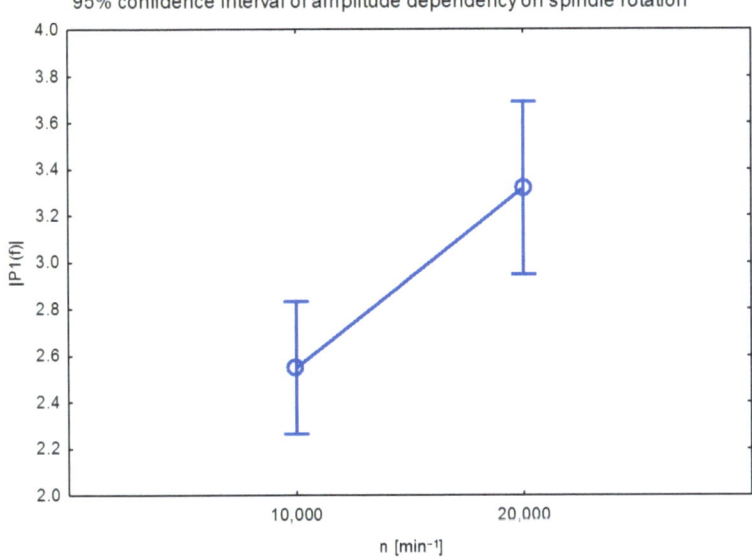

Figure 9. Box plot of 95% confidence interval of amplitude dependency on spindle rotation.

Figure 10 shows the dependency of amplitudes when changing the spindle rotations for different materials. When milling particleboard, amplitudes were lower than in milling MDF board.

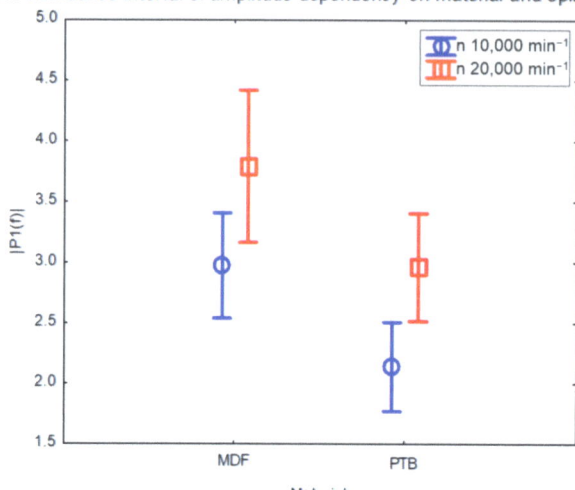

Figure 10. Box plot of 95% confidence interval of amplitude dependency on material and spindle rotation.

Figure 11 shows the dependency of amplitudes when changing feed rate for different experimental materials. As the feed rate increased, amplitudes increased at milling MDF boards bud decreased at milling particleboards.

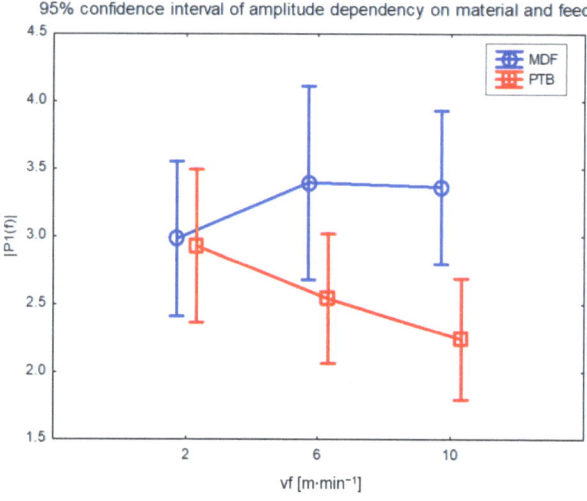

Figure 11. Box plot of 95% confidence interval of amplitude dependency on material and feed rate.

Figure 12 shows the dependency of amplitudes for different tools and experimental materials. The razor mill produced higher amplitudes of vibrations than the spiral mill at both experimental materials.

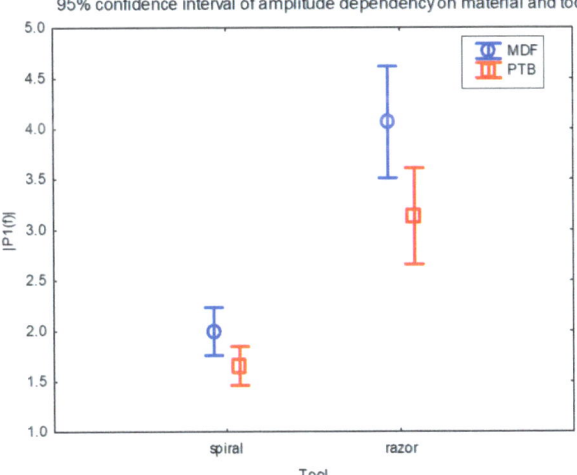

Figure 12. Box plot of 95% confidence interval of amplitude dependency on material and tool.

To verify the significance of differing parameters affecting amplitudes, the Duncan test was performed. A significant difference was found between tools, as shown in Table 6, spindle rotations are shown in Table 7, and materials are shown in Table 8. The change in feed rate, as shown in Table 9, did not impact amplitudes significantly.

Table 6. Duncan test of spindle rotation effect on amplitudes.

No.	n [min^{-1}]	Ampl. Mean	1	2
1	10,000	2.54		0.000842
2	20,000	3.32	0.000842	

Table 7. Duncan test of tool effect on amplitudes.

No.	Tool	Ampl. Mean	1	2
1	spiral	1.80		0.000009
2	razor	3.59	0.000009	

Table 8. Duncan test of material effect on amplitudes.

No.	Material	Ampl. Mean	1	2
1	MDF	3.28		0.000540
2	PTB	2.49	0.000540	

Table 9. Duncan test of feed rate effect on amplitudes.

No.	v_f [m·min^{-1}]	Ampl. Mean	1	2	3
1	2	2.96		0.899389	0.490981
2	6	2.92	0.899389		0.540870
3	10	2.75	0.490981	0.540870	

4. Discussion

The fast Fourier transform brought several results. First, as shown in Table 5, were dominant frequencies. The first peaks were observed at 165.93–166.13 Hz. The next peaks were observed at 332.26 Hz and 498.2 Hz. These, with rest of peaks, presented multiples of the first value. In some cases, multiples are not seen in Table 5. However, these were still present, but with maxima smaller than 0.5;0 they were filtered out. Similar results were observed through use of all parameters. Therefore, it seems the main vibrations were caused by the rotating tool.

This theorem would be supported by the maxima summary shown in Figure 8. When maxima from spindle rotation n = 10,000 min^{-1} are compared, the vibrations in x axis seem smaller or equal to axis z. As expected, the spiral mill produced higher results mainly in the z axis. With higher cutting speed n = 20,000 min^{-1}, the use of spiral mill produced larger maxima, which mostly seemed equally high in both x and z axis.

When comparing results from variance analysis, almost all parameter changes significantly affected the results. In Figure 9, the results show that the spindle speed had an effect on amplitude overall, with values in range (2.28–2.85) for n = 10,000 min^{-1} and (2.95–3.7) for n = 20,000 min^{-1}. The significance of this result is shown in Table 6. When different material was also considered, as shown in Figure 10, the results were similar—amplitude ranges were higher with higher cutting speed. Ranges were overall smaller with MDF, when compared to PTB.

The effect of the varying feed rate, as shown in Figure 11, did not seem as clear. While it could be stated for particleboard that a higher feed rate lowered maximum amplitudes, the same cannot be deduced for MDF. The result reached for PTB also seemed to contradict research where a lower feed rate resulted in higher surface quality [7]. The uncertainty of this result can be also seen in the results of the post hoc test shown in Table 9. This might be due to varying density of boards, where a larger number of specimens would bring a more definite result.

Lastly, as shown in Figure 12, the tool impacts were compared. In both materials, the use of spiral mill caused smaller amplitude ranges—(1.75–2.25) for MDF and (1.45–1.8) for PTB. The use of razor mill resulted in higher ranges (3.5–4.6) for MDF and (2.7–3.6) for PTB. These results were also confirmed by the Duncan test, as shown in Table 7.

5. Conclusions

The main goal of our research was to determine whether vibration monitoring on a gripper could be used for adaptive control, and therefore, if changes during the milling process would be detectable. To answer this question, tests were conducted to find if changes of parameters would produce differing measurements. The fast Fourier transform was used to process measured signals.

The most dominant frequencies (133 Hz and higher multiples) seemed to originate from tool cutting into material. When single-sided spectrum maxima were compared, the tool, spindle rotation, and feed rate seemed to affect the result.

The variance analysis and Duncan tests revealed a significant effect of tool, material, and spindle rotations. The feed rate analysis did not show conclusive results where the probability of similarity was over 5%.

Overall, vibrations measured at pneumatic gripper processed with fast Fourier analysis seemed to be sufficient as a potential signal for adaptive control during milling, similarly to [22,23]. As in [23], the roughness of specimens should be measured and paired with vibrations in a further study, to provide data for adaptive control model.

The results and procedures of this study will also serve for measurements of amplitude spectrum during nesting milling strategies in ongoing research. The limitation of the investigated measuring and processing system was that if the signal is to be evaluated in real time, it will require higher computational demands. The authors also intend to use this system in the development of a smart pneumatic gripper for the woodworking industry, where the modified procedures described in the article can be used.

Author Contributions: Conceptualization, P.K. (Peter Koleda) and R.K.; methodology, P.K. (Peter Koleda); software, Á.H.; validation, P.K. (Peter Koleda); formal analysis, Á.H.; investigation, P.K. (Peter Koleda) and P.K. (Pavol Koleda); resources, R.K.; data curation, Á.H.; writing—original draft preparation, Á.H.; writing—review and editing, P.K. (Peter Koleda); visualization, Á.H.; supervision, P.K. (Peter Koleda); project administration, P.K. (Peter Koleda); funding acquisition, P.K. (Peter Koleda). All authors have read and agreed to the published version of the manuscript.

Funding: This research was funded by "APVV-20-0403 FMA analysis of potential signals suitable for adaptive control of nesting strategies for milling wood-based agglomerates" and project VEGA 1/0791/21 "Research of non-contact method of analysis of small and dust particles arising in the production process with a prediction of negative effects of dust particles".

Institutional Review Board Statement: Not applicable.

Informed Consent Statement: Not applicable.

Data Availability Statement: Not applicable.

Acknowledgments: The authors would like to thank the Operational Program Integrated Infrastructure for the project: National Infrastructure for Supporting Technology Transfer in Slovakia II-NITT SK II, co-financed by the European Regional Development Fund.

Conflicts of Interest: The authors declare no conflict of interest.

Nomenclature

CNC	Computer Numerical Control
DC	Direct Current
FFT	Fast Fourier Transform
FMA	Failure Mode Analysis
HM tools	Hard Metal tools
ICEEMDAN	Improved Complete Ensemble Empirical Mode Decomposition with Adaptive Noise
MDF	Medium-Density Fiberboard
MEMS	Micro Electronic Mechanic System
N	revolutions (min^{-1})
PTB	Particleboard
WPT	Wavelet Packet Transform
v_f	feed rate (m·min^{-1})

References

1. Bendikiene, R.; Keturakis, G. The influence of technical characteristics of wood milling tools on its wear performance. *J. Wood Sci.* **2017**, *63*, 606. [CrossRef]
2. Kminiak, R.; Siklienka, M.; Šustek, J. Impact of tool wear on the quality of the surface in routing of MDF boards by milling machines with reversible blades. *Acta Fac. Xylologiae* **2016**, *58*, 89–100. [CrossRef]
3. Oláh, J.; Aburumman, N.; Popp, J.; Khan, M.A.; Haddad, H.; Kitukutha, N. Impact of Industry 4.0 on Environmental Sustainability. *Sustainability* **2020**, *12*, 4674. [CrossRef]
4. Kalsoom, T.; Ramzan, N.; Ahmed, S.; Ur-Rehman, M. Advances in Sensor Technologies in the Era of Smart Factory and Industry 4.0. *Sensors* **2020**, *20*, 6783. [CrossRef]
5. Cihad Bal, B.; Gündeş, Z. Surface Roughness of Medium-Density Fiberboard Processed with CNC Machine. *Measurement* **2019**, *153*, 107421.
6. İşleyen, Ü.; Karamanoğlu, M. The Influence of Machining Parameters on Surface Roughness of MDF in Milling Operation. *BioResources* **2019**, *14*, 3266–3277. [CrossRef]
7. De Deus, P.R.; Alves, M.C.S.; Vieira, F.H.A. The quality of MDF workpieces machined in CNC milling machine in cutting speeds, feedrate, and depth of cut. *Meccanica* **2015**, *50*, 2899–2906. [CrossRef]
8. Bal, B.; Dumanoğlu, F. Surface Roughness and Processing Time of a Medium Density Fiberboard Cabinet Door Processed via CNC Router, and the Energy Consumption of the CNC Router. *BioResources* **2019**, *14*, 9500–9508. [CrossRef]
9. Koc, K.H.; Erdinler, E.S.; Hazir, E.; Öztürk, E. Effect of CNC application parameters on wooden surface quality. *Measurement* **2017**, *107*, 12–18. [CrossRef]
10. Davim, J.P.; Clemente, V.C.; Silva, S. Surface roughness aspects in milling MDF (medium density fibreboard). *Int. J. Adv. Manuf. Technol.* **2009**, *40*, 49–55. [CrossRef]
11. Sütcü, A.; Karagöz, Ü. Effect of machining parameters on surface quality after face milling of MDF. *Wood Res.* **2012**, *57*, 231–240.

12. Kovatchev, G.; Atanasov, V. Determination of vibration during longitudinal milling of wood-based materials. *Acta Facultatis Xylologiae Zvolen Res. Publica Slovaca* **2021**, *63*, 85–92.
13. Sun, Z.; Geng, D.; Zheng, W.; Liu, Y.; Liu, L.; Ying, E.; Jiang, X.; Zhang, D. An innovative study on high-performance milling of carbon fiber reinforced plastic by combining ultrasonic vibration assistance and optimized tool structures. *J. Mater. Res. Technol.* **2023**, *22*, 2131. [CrossRef]
14. Pałubicki, B. Cutting forces in peripheral Up-milling of particleboard. *Materials* **2021**, *14*, 2208. [CrossRef] [PubMed]
15. Akbulut, T.; Ayrilmis, N. Effect of compression wood on surface roughness and surface absorption of medium density fiberboard. *Silva Fennica* **2006**, *40*, 161. [CrossRef]
16. Prakash, S.; Mercy, J.L.; Goswami, K. A systemic approach for evaluating surface roughness parameters during drilling of medium density fiberboard using Taguchi method. *Indian J. Sci. Technol.* **2014**, *7*, 1888–1894. [CrossRef]
17. Goli, G.; Curti, R.; Marcon, B.; Scippa, A.; Campatelli, G.; Furferi, R.; Denaud, L. Specific cutting forces of isotropic and orthotropic engineered wood products by round shape machining. *Materials* **2018**, *11*, 2575. [CrossRef]
18. Curti, R.; Marcon, B.; Furferi, R.; Denaud, L.; Goli, G. Specific cutting coefficients at different grain orientations determined during real machining operations. Procedings of the 24th International Wood Machining Seminar, Corvallis, OR, USA, 25–30 August 2019; pp. 53–62.
19. Kusuma, N.; Agrawal, M.; Shashikumar, P.V. Investigation on the influence of cutting parameters on Machine tool Vibration & Surface finish using MEMS Accelerometer in high precision CNC milling machine. *AIMTDR* **2014**, *375*, 1–6.
20. Bhogal, S.S.; Sindhu, C.; Dhami, S.S.; Pabla, B.S. Minimization of surface roughness and tool vibration in CNC milling operation. *J. Optim.* **2015**, *2015*, 192030. [CrossRef]
21. Liu, Y.; Zhang, D.; Geng, D.; Shao, Z.; Zhou, Z.; Sun, Z.; Jiang, Y.; Jiang, X. Ironing effect on surface integrity and fatigue behavior during ultrasonic peening drilling of Ti-6Al-4 V. *Chin. J. Aeronaut.* **2022**, in press. [CrossRef]
22. Asilturk, I. On-line surface roughness recognition system by vibration monitoring in CNC turning using adaptive neuro-fuzzy inference system (ANFIS). *Int. J. Phys. Sci.* **2011**, *6*, 5353–5360.
23. Plaza, E.G.; López, P.N.; González, E.B. Efficiency of vibration signal feature extraction for surface finish monitoring in CNC machining. *J. Manuf. Process.* **2019**, *44*, 145–157. [CrossRef]
24. Plaza, E.G.; López, P.N. Surface roughness monitoring by singular spectrum analysis of vibration signals. *Mech. Syst. Signal Process.* **2017**, *84*, 516–530. [CrossRef]
25. Misaka, T.; Herwan, J.; Ryabov, O.; Kano, S.; Sawada, H.; Kasashima, N.; Furukawa, Y. Prediction of surface roughness in CNC turning by model-assisted response surface method. *Precis. Eng.* **2020**, *62*, 196–203. [CrossRef]
26. Bouhalais, M.L.; Nouioua, M. The analysis of tool vibration signals by spectral kurtosis and ICEEMDAN modes energy for insert wear monitoring in turning operation. *Int. J. Adv. Manuf. Technol.* **2021**, *115*, 2989–3001. [CrossRef]
27. Lin, Y.-C.; Wu, K.-D.; Shih, W.-C.; Hsu, P.-K.; Hung, J.-P. Prediction of surface roughness based on cutting parameters and machining vibration in end milling using regression method and artificial neural network. *Appl. Sci.* **2020**, *10*, 3941. [CrossRef]
28. Lin, W.-J.; Lo, S.-H.; Young, H.-T.; Hung, C.-L. Evaluation of deep learning neural networks for surface roughness prediction using vibration signal analysis. *Appl. Sci.* **2019**, *9*, 1462. [CrossRef]
29. Nasir, V.; Sassani, V. A review on deep learning in machining and tool monitoring: Methods, opportunities, and challenges. *Int. J. Adv. Manuf. Technol.* **2021**, *115*, 2683–2709. [CrossRef]
30. Siklienka, M.; Janda, P.; Jankech, A. The influence of milling heads on the quality of created surface. *Acta Fac. Xylologiae Zvolen Res Publica Slovaca* **2016**, *58*, 81.
31. Vitchev, P. Evaluation of the surface quality of the processed wood material depending on the construction of the wood milling tool. *Acta Fac. Xylologiae Zvolen* **2019**, *61*, 81–90.
32. Płodzień, M.; Żyłka, Ł.; Sułkowicz, P.; Żak, K.; Wojciechowski, S. High-performance face milling of 42CrMo4 steel: Influence of entering angle on the measured surface roughness, cutting force and vibration amplitude. *Materials* **2021**, *14*, 2196. [CrossRef] [PubMed]
33. Cochran, W.T.; Cooley, J.W.; Favin, D.L.; Helms, H.D.; Kaenel, R.A.; Lang, W.W.; Maling, G.C.; Nelson, D.E.; Rader, C.M.; Welch, P.D. What is the fast Fourier transform? *Proc. IEEE* **1967**, *55*, 1664–1674. [CrossRef]
34. Shrivastava, Y.; Singh, B. A comparative study of EMD and EEMD approaches for identifying chatter frequency in CNC turning. *Eur. J. Mech.-A/Solids* **2019**, *73*, 381–393. [CrossRef]
35. Palani, S.; Natarajan, U. Prediction of surface roughness in CNC end milling by machine vision system using artificial neural network based on 2D Fourier transform. *Int. J. Adv. Manuf. Technol.* **2011**, *54*, 1033–1042. [CrossRef]
36. Wang, S.-M.; Ho, C.-D.; Tsai, P.-C.; Yen, C. Study of an efficient real-time monitoring and control system for BUE and cutter breakage for CNC machine tools. *Int. J. Precis. Eng. Manuf.* **2014**, *15*, 1109–1115. [CrossRef]
37. Ward, R.; Sencer, B.; Panoutsos, G.; Ozturk, E. On-The-Fly CNC interpolation using frequency-domain FFT-based filtering. *Procedia CIRP* **2022**, *107*, 1571–1576. [CrossRef]
38. Fast Fourier Transform. Available online: https://www.mathworks.com/help/Matlab/ref/fft.html (accessed on 1 March 2023).
39. Kowaluk, G.; Szymanski, W.; Palubicki, B.; Beer, P. Examination of tools of different materials edge geometry for MDF milling. *Eur. J. Wood Wood Prod.* **2009**, *67*, 173–176. [CrossRef]
40. Laszewicz, K.; Górski, J.; Wilkowski, J. Long-term accuracy of MDF milling process—Development of adaptive control system corresponding to progression of tool wear. *Eur. J. Wood Wood Prod.* **2013**, *71*, 383–385. [CrossRef]

41. Zbieć, M. Application of Neural Network in Simple Tool Wear Monitoring and Indentification System in MDF Milling. *Drv. Ind.* **2011**, *62*, 43–54. [CrossRef]
42. Laszewicz, K.; Górski, J. Practical use of control charts as a quality control tool (size precision) in the process of milling of MDF boards. *Ann. Warsaw Univ. Life Sci.-SGGW For. Wood Technol.* **2012**, *78*, 206–211.
43. Hesselbach, J.; Hoffmeister, H.W.; Schuller, B.C.; Loeis, K. Development of an active clamping system for noise and vibration reduction. *CIRP Ann.* **2010**, *59*, 395–398. [CrossRef]
44. Melnik, Y.A.; Kozochkin, M.P.; Porvatov, A.N.; Okunkova, A.A. On adaptive control for electrical discharge machining using vibroacoustic emission. *Technologies* **2018**, *6*, 96. [CrossRef]
45. Nahornyi, V.; Panda, A.; Valíček, J.; Harničárová, M.; Kušnerová, M.; Pandová, I.; Legutko, S.; Palková, Z.; Lukáč, O. Method of Using the Correlation between the Surface Roughness of Metallic Materials and the Sound Generated during the Controlled Machining Process. *Materials* **2022**, *15*, 823. [CrossRef]
46. Iskra, P.; Tanaka, C. The influence of wood fiber direction, feed rate, and cutting width on sound intensity during routing. *Eur. J. Wood Wood Prod.* **2005**, *63*, 167–172. [CrossRef]
47. Pantaleo, A.; Pellerano, A. Assessment of wood particleboards milling by means of energy consumption tests. *Wood Mater. Sci. Eng.* **2014**, *9*, 193–201. [CrossRef]
48. Licow, R.; Chuchala, D.; Deja, M.; Orlowski, K.A.; Taube, P. Effect of pine impregnation and feed speed on sound level and cutting power in wood sawing. *J. Clean. Prod.* **2020**, *272*, 122833. [CrossRef]
49. EN 14322; Wood-Based Panels-Melamine Faced Board for Interior Uses—Definition, Requirements and Classification. European Committee for Standardization: Brussels, Belgium, 2017.
50. EN 312-2; Particleboards-Specifications-Part 2: Requirements for General Purpose Boards for Use in Dry Conditions. European Committee for Standardization: Brussels, Belgium, 1996.
51. EN ISO 12460-5; Determination of Formaldehyde Release-Part 5: Extraction Method (Called the Perforator Method). European Committee for Standardization: Brussels, Belgium, 2016.
52. Three Axis Accelerometer Kit—Quick Start Guide. Available online: https://www.picotech.com/download/manuals/ThreeAxisAccelerometerKitQSG.pdf (accessed on 28 February 2023).
53. Hortobágyi, Á.; Koleda, P.; Kminiak, R. Workpiece gripper vibration measurement for nesting milling. *Trieskové A Beztrieskové Obrábanie Dreva* **2022**, *13*, 19–24.

Disclaimer/Publisher's Note: The statements, opinions and data contained in all publications are solely those of the individual author(s) and contributor(s) and not of MDPI and/or the editor(s). MDPI and/or the editor(s) disclaim responsibility for any injury to people or property resulting from any ideas, methods, instructions or products referred to in the content.

Article

Impact of Wood Structure Variability on the Surface Roughness of Chestnut Wood

Marina Chavenetidou [1] and Vasiliki Kamperidou [2,*]

[1] Laboratory of Wood Utilization, Department of Harvesting and Technology of Forest Products, Forestry and Natural Environment Faculty, Aristotle University of Thessaloniki, 54124 Thessaloniki, Greece; mchavene@for.auth.gr

[2] Laboratory of Wood Technology, Department of Harvesting and Technology of Forest Products, Forestry and Natural Environment Faculty, Aristotle University of Thessaloniki, 54124 Thessaloniki, Greece

* Correspondence: vkamperi@for.auth.gr; Tel.: +30-2310992747

Citation: Chavenetidou, M.; Kamperidou, V. Impact of Wood Structure Variability on the Surface Roughness of Chestnut Wood. *Appl. Sci.* **2024**, *14*, 6326. https://doi.org/10.3390/app14146326

Academic Editors: Martin Kučerka, Alena Očkajová and Richard Kminiak

Received: 2 July 2024
Revised: 15 July 2024
Accepted: 19 July 2024
Published: 20 July 2024

Copyright: © 2024 by the authors. Licensee MDPI, Basel, Switzerland. This article is an open access article distributed under the terms and conditions of the Creative Commons Attribution (CC BY) license (https://creativecommons.org/licenses/by/4.0/).

Abstract: Wood constitutes a unique and valuable material that has been used from ancient times until nowadays in a wide variety of applications, in which the surface quality of wood often constitutes a critical factor. In this study, the influence of different wood areas and therefore, of different anatomical characteristic areas of chestnut wood (*Castanea sativa* Mill.) on the surface quality, was thoroughly studied, in terms of surface roughness. Five different chestnut tree trunks were harvested, from which five different disks were obtained corresponding to five different trunk heights. Surface roughness was measured on these disks on the transverse, radial, and tangential planes, on the areas of sapwood and heartwood, measuring the roughness in each point both vertically and in parallel to the wood grain. The results revealed that the examined roughness indexes (Ra, Rz, Rq) follow a parallel path to one another. In the case of all surfaces (transverse, radial, tangential) of the disks examined, when the measurement was implemented perpendicularly to the wood grain, a significantly higher roughness was recorded, compared to the wood grain measurements being implemented in parallel with the wood grain. Significant differences between heartwood and sapwood roughness were not demonstrated, although sapwood often appeared to exhibit a higher surface roughness than heartwood sites. Among the roughness values of the three different surfaces, the highest roughness in the vertical-to-wood-grain measurements was recorded by tangential surfaces, with slightly lower values on the transverse surfaces and the lowest roughness on radial surfaces. Meanwhile, for the measurements in parallel with the wood grain, the transverse surfaces presented significantly higher roughness values compared to the tangential and radial surfaces. Significant roughness differences were not detected among the surfaces at different trunk heights. Although, significant differences in roughness were recorded among different trees, it was observed that all the studied trees align with the identified and described within-tree trends.

Keywords: heartwood; quality; radial; roughness; sapwood; structure; surface; tangential; texture; wood

1. Introduction

The quality of wood surfaces is highly crucial for the manufacturing of qualitative wood-based products and structures. More specifically, the surface roughness is a matter of great interest for numerous applications of wood in small- or bigger-dimension structures (furniture, floors, frames, paneling, table-tops, etc.), defining their appearance, texture, aesthetics, and user-generated sensation, among others [1,2]. Furthermore, low surface roughness values, with regard to several wood species and used adhesives, have demonstrated higher shear bonding strength results [3]. Wood surface quality depends mainly on the wood structure and the implementation of wood mechanical processing procedures, such as cutting, sanding, finishing, painting, curving, applying preservation methods, adhesives, coatings, other substance layers, etc. [4–7].

As a biological anisotropic material with a multidimensional surface area, wood texture is closely related to its structural and anatomical features (fibers, pores, tracheids, rays, etc.) and its formation and appearance are affected by various factors, either environmental or genetic. Different wood species, either softwood or hardwood, of different origin appear to show variability in structure and properties [8,9], while even in the same trunk, large differences in wood structure can be detected and, therefore, in wood surface quality as well. Characteristics, such as the growth ring width, their homogeneity and appearance, cell wall thickness, cell type composition and wood density, heartwood and sapwood ratio, earlywood and latewood ratio, wood moisture content, surface planes (tangential, radial, transverse), etc., undoubtedly influence the quality of a wood surface, forming numerous geometric peaks and valleys [10,11]. As reported by Sadoh and Nakato [12], the diffuse-porous wood species present lower surface roughness values than the ring-porous wood species. Örs and Gürleyen [13] reported that compared to the radial planes, the tangential planes presented higher surface quality. Lower surface roughness has been recorded in latewood than the earlywood [14], probably because of the thicker cell walls and the higher density of latewood. The factors of ring width, wood density, and ring angle also seem to highly influence the surface roughness [15].

Moreover, the mechanical processing using different cutting machines and tool influences also affect the wood surface roughness. The final quality of a wood surface is strongly dependent on cutting procedure kinematics, the feed speed, the rake angle [14], and other preparation processes such as sanding and finishing [16,17], with machining defects such as fuzzy, torn, or raised grain being correlated with high surface roughness. According to the literature, slow-feeding wood planes outperformed high-feeding wood planes in terms of surface roughness [18]. The level of maintenance of these machines and the cutting means they carry, the storage conditions of the wood, the moisture content fluctuation until the final use, and the subsequent dimensional stability of the wood, all constitute factors that usually provide a totally different wood substrate in terms of roughness [7,19].

Therefore, roughness reflects the combined effect of several different factors simultaneously interacting and perhaps that is one of the reasons why it has not been thoroughly comprehended so far, although it constitutes a property that significantly affects the utilization degree of wood in several applications. It has also been revealed that surface quality, and more specifically, surface roughness, is closely related to the duration and service life of timber and the respective structures it participates in [20]. Smoother surfaces prove to be more resistant to stress and wear [14]. Since high roughness corresponds to discontinuities in the wood tissue, it is expected and inevitable that this is also associated with the retention of a higher moisture content, the higher potential of wood biological damage, wood degradation by the action of microorganisms, a deterioration in wood substrate, etc. Therefore, the smoothness of wood surfaces has been associated with a longer service-life duration of wood and wood structures.

Currently, empirical procedures and models have been applied, using several surface roughness indexes of *Ra*, *Rz*, *Rq*, *Rk* and *Rap*, among others [21]. However, the most commonly used surface roughness indexes being recorded, in order to define wood surface roughness, are as follows: *Ra*, which corresponds to the average of the values of the roughness profile; *Rz*, the mean value of the roughness depths of different sampling lengths; and *Rq*, the largest roughness depth width. Determination methods can be applied either with or without contact with the wood surface. Contact methods employ a stylus tip, pneumatic methods, and tactile sensation [22]. A quite commonly used method is the use of a profilometer bearing a diamond stylus, which runs a path on the surface and records the surface roughness in different directions of wood grain [20,22,23].

The wood of chestnut is regarded as being classified among the most valuable timber species of Europe and presents a great range of uses and applications in the form of round timber, technical sawn wood, floors, furniture, high-value items, etc. [24,25]. It is a ring-porous hardwood species with earlywood vessels that are of significantly higher diameter than those of latewood, presenting a clear ring arrangement [25]. Especially when sapwood is being transformed into heartwood, the earlywood vessels are usually full of tyloses, while latewood vessels are polygonal and are found in groups generating a flame-like design. This species' rays are more often uniseriate, rarely biseriate, and of different heights [8]. The parenchyma is mostly apotracheal and rarely paratracheal in places [8,25,26]. Chestnut wood is considered to be of medium density, approximately 0.57–0.63 g/cm^3, and in general, density is strongly related to ring width. Especially for ring-porous hardwoods such as chestnut, growth rings of high width tend to demonstrate higher values of density, due to the fact that earlywood remains more or less stable, while the increase in growth ring width is attributed to the increase in the latewood part. However, density is also slightly affected by the cambium age and appears to decrease as the tree ages [27,28]. The outdoors exposure of wood, where intensive changes in environmental conditions take place, concerning relative moisture content, atmospheric precipitation, UV radiation, etc., causes chestnut wood to gradually deteriorate and discoloration of the surfaces occurs, regardless of whether the wood has been coated with mild hydrophobic solutions or not, which subsequently deteriorate the quality and roughness of wood surfaces [29].

According to the literature, only Sutcu and Karagoz [30] have dealt with surface quality of chestnut among other species. More specifically, they investigated how machining conditions (feed rate, spindle speed, step-over, axial depth, etc.) affected the roughness of chestnut, beech, and walnut specimens, concluding that the wood roughness was greatly influenced by the factors of cut depth, feed rate, and spindle speed. Therefore, a great lack of research measurements and data have been identified in the literature with regard to the surface roughness of chestnut wood and the factors influencing it, although chestnut constitutes such an important timber species, frequently used in applications where roughness is a crucial parameter. To the best of our knowledge, in the literature, there is no study providing surface roughness data of chestnut wood, concerning the potential surface roughness vertical variability, the variability among wood sections horizontally (sapwood, heartwood)/planes (transverse, radial, tangential), or findings/information about any potential correlation between structural characteristics and the roughness of chestnut wood.

Therefore, the aims of the current study are to thoroughly examine the surface quality of chestnut wood, in terms of surface roughness, examining the potential differences in the three wood surface planes (transverse, radial, and tangential), in the different areas of sapwood and heartwood, and to examine the potential variability "among different chestnut trees" and "among different trunk heights" of surface roughness. In addition, "between different direction measurements" (vertically/in parallel with the wood grain) will also be examined in order to conduct a thorough roughness characterization of chestnut wood. Potential correlations between wood structure variables and surface roughness are going to be examined. The implementation of the current research is anticipated to provide an insight into the scientific field of wood surface quality and roughness, as well as the impact of the anatomical characteristics of chestnut wood material on wood roughness and its rational utilization in various applications.

2. Materials and Methods

2.1. Sample Preparation

For the purposes of this study, five chestnut (*Castanea sativa* Mill.) trees were harvested from a coppice forest in Sithonia Peninsula (Chalkidiki, Greece). The trees were as straight as possible, without any apparent defects. The trees were aged 25–27 years and their diameter ranged from 19.1 cm to 24 cm. Tree trunk disks of 3 cm thickness were obtained for approximately every 1 m of height from the tree trunk base (starting from the height of 1 m) to the top, taking a total of 5 disks per trunk (25 disks in total).

The disks were transferred to laboratory infrastructure and conditioned in a closed chamber under stable conditions (60% relative humidity, 20 ± 3 °C) until a constant weight was achieved. The moisture content of the disk wood was recorded according to the ISO

13061-1 [31] standard to be 7.6–8.5%. All the measurements were performed on the different surface planes of these disks (Figure 1).

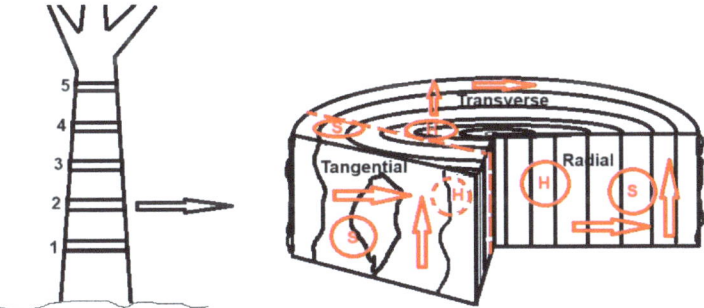

Figure 1. Configuration of sampling method of the 5 disks per tree (on the **left**) and areas of interest for measurement (on the **right**). The transverse surface/plane is perpendicular to the stem longitudinal axis; the radial surface is oriented along the direction of a ray of the circumference described by the stem; the tangential surface is perpendicular to the direction of a ray of the circumference described by the stem. The arrows depict the direction of measurement (parallel with or vertical to the wood grain); "H" corresponds to heartwood and "S" to sapwood area.

First of all, the disks were further cut using a band saw (Figure 2A) in order to form samples of smaller dimensions bearing clear transverse, radial, and tangential surfaces concerning the heartwood and sapwood areas of wood (Figure 2B). Afterwards, the samples were code marked, sanded using 80-grit sandpaper on a sanding machine under the same processing conditions (TC-US400, Einhell, Germany) (Figure 2C), and a polishing technique was applied to all the samples to ensure the comparability and reliability of the results. In general, if high-grit sanding is applied during surface preparation, roughness would be only affected by wood structure and ingenuine properties. However, it is difficult to relate wood anatomy to roughness, since sanding is a procedure that may be conducted in different conditions and with different means. For instance, when high-speed cutting is performed, softwood species are strongly affected [4,32]. Therefore, in the current study, only one and the same person/operator implemented all the sample sanding processes, applying the same sanding process on the same device, spending the same time on each wood sample (approximately 2 min/surface). All the specimens were visually and empirically examined and determined to have been appropriately sanded.

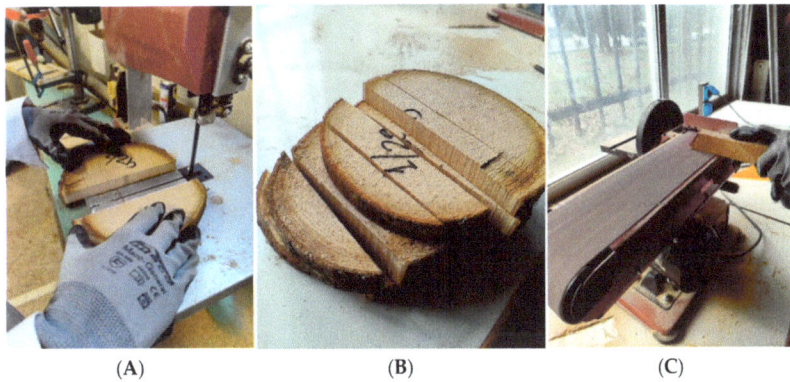

Figure 2. Chestnut wood disk/specimen preparation ((**A**) cutting, (**B**) marking, (**C**) sanding processes).

2.2. Physical and Chemical Property Assessment

Prior to the roughness measurement, the dry density and maximum moisture content of the samples (maximum moisture that can be absorbed/retained) were determined according to processes described by Tsoumis [8], applying the equation of $R_0 = M_0/V_0$, where R_0 corresponds to the dry density (g/cm^3), M_0 is the dry mass (g), and V_0 is the dry volume (cm^3). Specifically, samples were formed in stripes without bark (beginning from the diameter line of the trunk cross-section) and then, each stripe was split into pieces of approximately 2–5 annual rings. Dry volume of the pieces was estimated with the application of the water displacement method based on Archimede's principle (which states that a body immersed in a fluid is subjected to an upwards force equal to the weight of the displaced fluid. This is a first condition of equilibrium) and was the dry weight after heating for 24 h in an oven at 103 ± 2 °C. Maximum moisture content was determined by applying weight measurements before and after the kiln drying for 24 h and until the stabilization of the values. In most cases, measurements were conducted before and after the extraction of the samples with hot water, so as to evaluate the relation between the extractives presence and maximum moisture content and dry density and therefore, also with wood surface roughness.

The wood extractive content (of those extracted with boiling water, referring mainly to tannins, gum, sugars, coloring substances) of chestnut wood was also measured in the current research, applying the common lab process (100 °C, 3 h, in a thermal jacket and water vapor cooler system) [9].

2.3. Roughness Measurements

The measurements were conducted using a profilometer "Mitutoyo Surftest SJ-301" fine stylus, based on ISO 21920-2:2021 [33] methodology (Figure 3). The measurements were implemented on the prepared disks, at selected wood surface areas without any defect, both in parallel with and perpendicular to the grain, on transverse, radial, and tangential sections/planes of heartwood and sapwood areas. The applied methodology [33], as well as the instructions of the profilometer manufacturer and previous published studies [5,34,35] were followed. The parameters of roughness that were recorded were *Ra* (mean arithmetic profile deviation), *Rz* (mean peak to valley height), and *Rq* (maximum roughness), based on previous relevant studies dealing with wood surface quality. Approximately 15 measurements were obtained from each studied case (heartwood/sapwood, transverse/radial/tangential section, parallel/perpendicular to the grain).

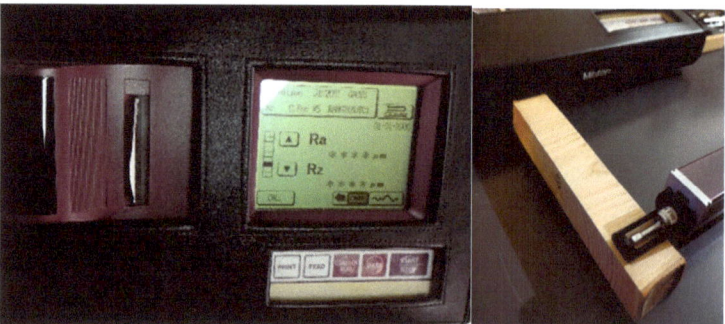

Figure 3. Profilometer device and view of surface roughness measurements on wood on the right.

The measurement points were chosen randomly on the surface of the samples, in order to ensure that both areas of earlywood or latewood in the three directions would be involved and the whole surface area of each sample would be covered. More specifically, a 2-dimensional rectangular sampling grid with points spaced 10 mm apart was placed over each of these created areas of the wood disk specimens. Points found at each intersection of the grid were marked on the surface and included in the measurements. Particular attention was paid to ensure the representativeness of the samples.

Additionally, prior to the process of roughness measuring, a calibration of the instrument preceded, and temperature conditions were approximately $20 \pm 3\,°C$ [7,34,36].

2.4. Statistical Analysis

The statistical analysis of the results was implemented, initially applying a one-way ANOVA, using the statistical package of SPSS (Statistics PASW), in order to identify which of the surface roughness values of these categories differed from one another in a statistically significant manner. Secondly, a two-way ANOVA was applied examining each of the variable pairs, in which significant differences were recorded. In this way, potential variable correlations were investigated and the impact of each of the different factors (independent variables) on the dependent variable, level of surface roughness (Ra), were assessed. All possible pairings and correlations among the variables were examined using the "enter" approach. Non-statistically significant variables were excluded from the analysis through the "stepwise" approach. The "stepwise" method, known for its strictness, permits only statistically significant variables with a meaningful effect on the dependent variable to be included in the multiple linear regression analysis. To detect potential autocorrelation in the residuals of the regression analysis, the Durbin Watson statistic was calculated. All statistical analyses were conducted with a significance level set at p-value = 0.05.

3. Results and Discussion

3.1. Physical and Chemical Properties Assessment

According to the results of this research, the average wood dry density was approximately $0.60\ g/cm^3$ and ranged from $0.57\ g/cm^3$ to $0.63\ g/cm^3$ (Table 1). The mean maximum moisture content of chestnut wood was measured to be 100.43%.

Table 1. Density (dry and basic), maximum moisture content values (%), and extractive content (%) of the chestnut wood studied.

Tree		Dry Density g/cm³	Basic Density g/cm³	Maximum MC %	Extractives %
1	x	0.565	0.526	110.236	13.208
	s±	0.022	0.022	7.683	3.179
	n	20	20	20	4
2	x	0.595	0.599	102.235	11.385
	s±	0.033	0.030	7.063	2.326
	n	23	23	23	4
3	x	0.599	0.561	100.696	14.133
	s±	0.019	0.0202	4.736	2.6441
	n	21	21	21	4
4	x	0.607	0.564	96.651	15.084
	s±	0.025	0.026	5.958	3.2797
	n	20	20	20	4
5	x	0.633	0.583	92.354	10.637
	s±	0.037	0.030	6.951	1.6802
	n	23	23	23	4
Mean	x	0.600	0.599	100.434	12.889

x: mean value, s±: standard deviation value, n: number of examined samples.

Extractive content, as well as other factors such as the presence of tyloses, especially in the heartwood area, are considered to be strongly related to increased wood density especially in the heartwood area, and as a result, to a decrease in surface roughness. The roughness and, in general, the surface quality of wood, is closely related to its density, as well as its anatomical characteristics and chemical composition [9]. More specifically, the higher the density of the wood, the smoother its surface, most of the time.

Therefore, as can be seen in Table 1, chestnut is a species with a high content of extractives and this is in line with medium to high values of its density, especially in the heartwood area [8,27].

3.2. Surface Roughness

The three examined roughness indexes, *Ra*, *Rz*, and *Rq*, follow a parallel path to one another, a trend that is also highlighted in the provided summary/clustering diagrams of the five trees (Figures 4–8). Based on this and the available literature, the *Ra* index was used in the statistical analyses of the current study to detect correlations among different wood structure factors.

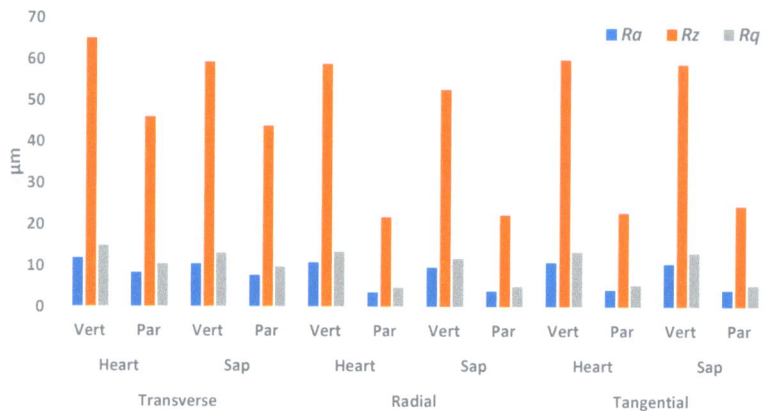

Figure 4. Summary bar plot presenting the surface roughness index values (*Ra*, *Rz*, *Rq* in µm) of tree 1 (including the 5 disks).

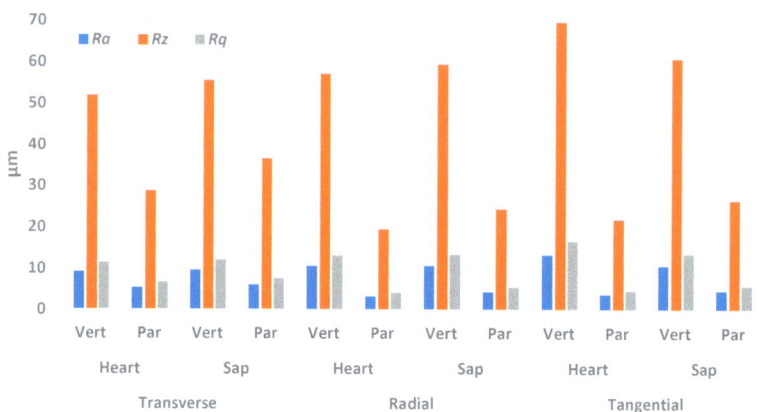

Figure 5. Summary bar plot presenting the surface roughness index values (*Ra*, *Rz*, *Rq* in µm) of tree 2 (including the 5 disks).

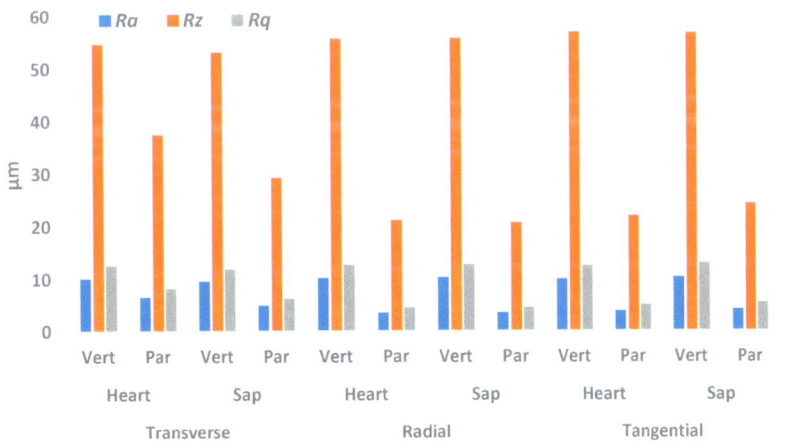

Figure 6. Summary bar plot presenting the surface roughness index values (*Ra*, *Rz*, *Rq* in μm) of tree 3 (including the 5 disks).

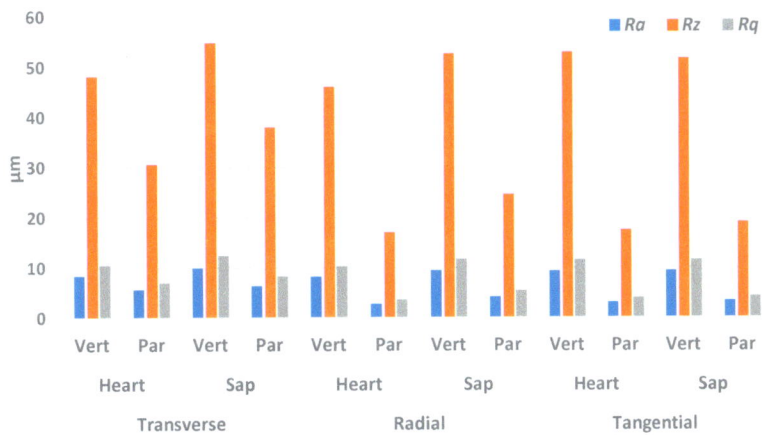

Figure 7. Summary bar plot presenting the surface roughness index values (*Ra*, *Rz*, *Rq* in μm) of tree 4 (including the 5 disks).

Chestnut wood, characterized by its medium value of density (approximately 0.57 g/cm^3) and relatively medium to high hardness values (approximately 41.07 N/mm^2 tangentially and 40.29 N/mm^2 radially), was expected to show medium surface roughness values [37–39]. In general, density and porosity are inversely correlated with each other; therefore, both highly affect smoothness, since high-density samples show lower roughness, though sometimes, also a higher resistance during the sanding process [38,39]. Fortino et al. [40] found a correlation of surface roughness with the hardness and scratch resistance of the wood, which applied in the presence of different moisture contents. They concluded that the effect of wood structures such as earlywood and latewood, and sapwood and heartwood were crucial factors. For instance, the sapwood zones in softwood species show a higher scratch resistance (compared to heartwood), with higher scratch resistance values recorded in the tangential direction. Finally, the scratch resistance is lower in earlywood locations compared to latewood.

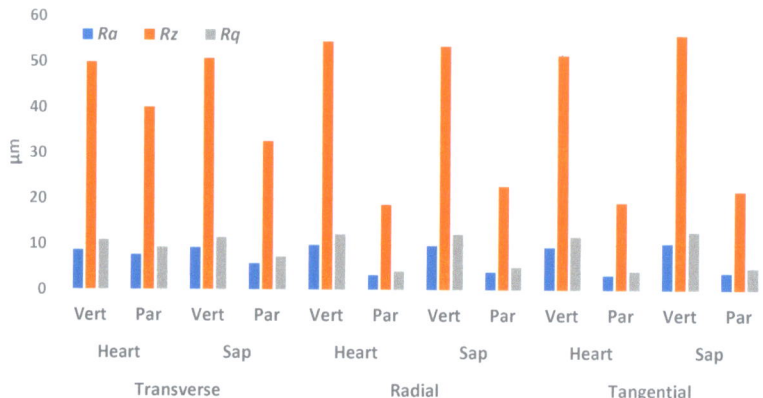

Figure 8. Summary bar plot presenting the surface roughness index values (*Ra*, *Rz*, *Rq* in μm) of tree 5 (including the 5 disks).

In the current study, statistically significant differences in roughness were not detected among the values of the different heights of the tree (among the different tree disks, 1–5, of each tree) when comparing the roughness values in the same variables (measurement direction, specimen's surface planes, areas of sapwood/heartwood). More specifically, according to the statistical analysis, the surface roughness (*Ra*) was affected by only 4.5% by the height of the specimen location in the trunk (different disks), presenting higher roughness in the higher tree heights. Statistically significant differences were recorded only between the lowest disks (1–2) and the highest disk (disk 5), proving that there is a slight difference in the surface roughness of the wood depending on the location of the wood specimen in the trunk longitudinally (trunk height). This could be probably attributed to the fact that in the lower tree heights, near the base of the tree, the wood is more mature and the density is higher due to the higher proportion of heartwood, among other factors.

The surface roughness, in general, was affected by 72.9% by the factors "Measurement Direction" (vertically/parallel), "Planes" (transverse, radial, tangential surfaces), and the factor "Area" (sapwood/heartwood) in combination. The interaction of the factor trunk "Height" with the abovementioned factors of "Direction", "Planes", and "Area" affected the variability of *Ra* only by 8%, providing more evidence that the impact of trunk height can be considered to be of lower significance.

Nevertheless, significant differences appeared among different trees (comparing the values of measurements of the respective variables of direction, plane types, disks, etc.), with "Tree 3" revealing the statistically significant and lowest roughness compared to the rest of the trees, while the described within-tree trends that were detected apply to each of the tree cases.

The results of the current study also revealed that in all surface planes (transverse, radial, tangential) of the examined disks, when the measurement was implemented vertically to the wood grain, a statistically significant higher roughness value was recorded, compared to the measurement implemented in parallel with the wood grain, which corresponds to statistically significant differences in all the studied cases. It is characteristic that 67.6% of the variability of roughness is being influenced by the factor of measurement "Direction" (orientation of vertically/parallel to wood grain). This tendency could be easily explained, taking into account that vertical to the wood grain, higher height differences are encountered due to earlywood–latewood transition zone areas, different growth rings, etc. Chestnut, as a ring-porous hardwood species, also demonstrates differences in cell wall thickness between the earlywood and latewood areas and as is widely accepted [41], the surface roughness is strongly associated with cell wall thickness. More specifically, latewood fibers present thicker cell walls than those of earlywood [8]. The presence of earlywood vessels that are characterized by a much higher cell diameter compared to latewood results in density differences in each of the growth rings [8].

In this study, statistically significant differences between heartwood and sapwood roughness values were not demonstrated, although sapwood appeared in some cases to exhibit slightly a higher surface roughness than the corresponding heartwood sites (referring to the same direction of roughness measurement). The wood tissue found in the heartwood part of the trunk consists of cells that have ceased to serve as part of the tree's conduit system and the cells have been filled with storage/healing substances, extracts, etc., presenting a slightly higher density.

The density of wood, in combination with its structural characteristics and chemical composition, are all strongly correlated with its roughness [9], with the higher density corresponding most of the time to smoother surfaces.

Among the roughness values of the three different surfaces examined on the disks, it was observed that the highest roughness values (*Ra*) were detected, in most disk cases, on the tangential surface of the disks, then on the transverse disks, and finally, the lowest *Ra* values were recorded on the radial surfaces (concerning the vertical-to-the-grain measurements). This finding could be attributed to the presence of radii on the radial surfaces that probably make the wood surface smooth, as well as to the fact that the tangential surfaces in the samples taken corresponded mostly to the sapwood part of the tree, and therefore contained a higher proportion of sapwood than heartwood, which probably contributed to an increase in the surface roughness. Additionally, concerning the vertical measurements of all the categories, statistically significant differences were not recorded among the different categories. Meanwhile,, regarding the parallel-to-the-grain measurements of all the categories, statistically significant differences were found, with the transverse surfaces of the examined wood specimens to reveal the highest roughness values (*Ra*) (Figure 9). Therefore, when the roughness measurements were conducted parallel to the wood grain, the highest roughness was observed in transverse surfaces, which corresponded to a statistically significant difference between transverse and the other two surface planes (tangential and radial), and these two did not differ significantly from one another. This much higher roughness recorded in transverse surfaces (when measured parallel to the wood grain) could be explained by the fact that chestnut is a ring-porous wood species and when the measurement of roughness is conducted in parallel orientation to the wood grain, there is a high potential for the measurement to be implemented alongside the earlywood area, which consists of cross-sectional cut vessels (of large diameter) [8] that would definitely increase the roughness of the surface.

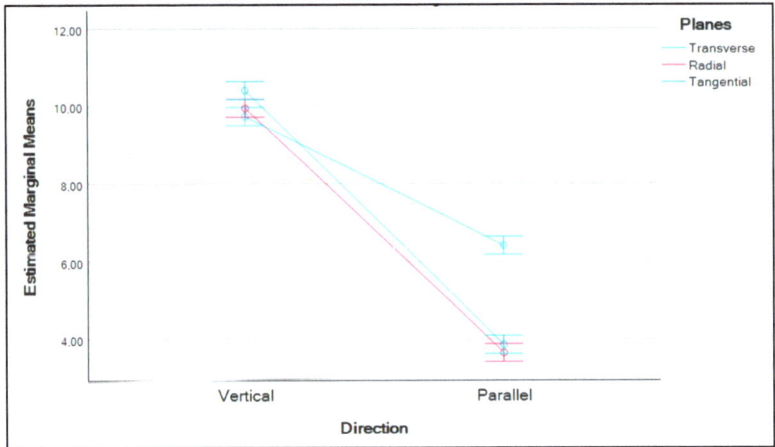

Figure 9. Estimated marginal means plot of surface roughness index of *Ra* (μm), recorded in measurements that were conducted in different orientations (vertically and parallel to the wood grain).

After the examination of all possible pairs of factors in terms of combination and interaction, we decided to apply a "decomposition" of the statistical analysis, abstracting the factor "Area" (heartwood–sapwood) due to the fact that it did not demonstrate a statistically significant impact. Moreover, the results of the remainder of the variable combinations revealed that the 67.6% of the variability of surface roughness (*Ra*) is being influenced by the factor of measurement "Direction" (vertically/parallel) and only 7.8% by the factor "Planes" (transverse, radial, tangential), proving that the "Planes" variable, although inducing statistically significant differences in roughness values, has quite a low impact on the surface roughness (Figure 10).

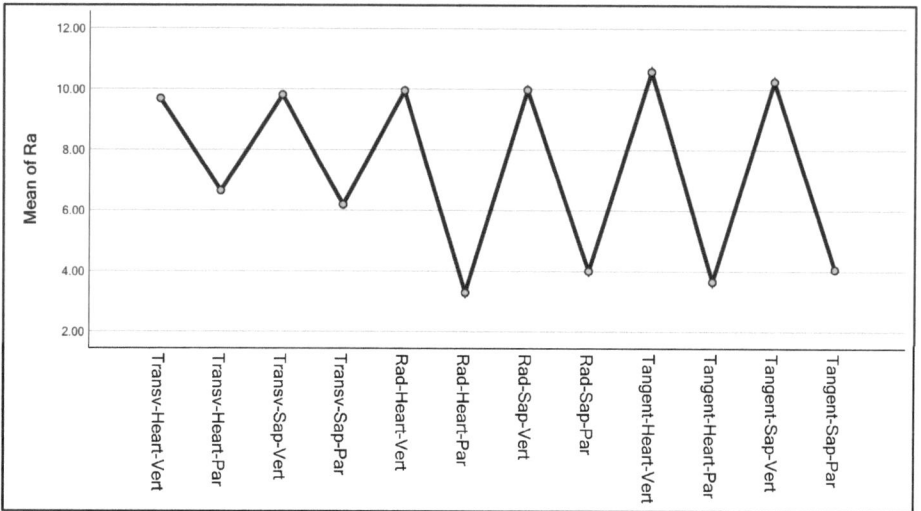

Figure 10. Clustering line chart of surface roughness index (*Ra*, μm) depending on the different planes (transverse, radial, tangential), area (sapwood, heartwood), and direction of measurement (vertically, parallel).

Additionally, regarding the transverse plane measurements, the *Ra* variability was found to be influenced by the factor measurement "Direction" (vertically/parallel) by 20.9% (Table 2). In the radial plane measurements, the impact of the measurement "Direction" factor on *Ra* was found to be 48%, while in tangential measurements, 50.7% of the *Ra* variability was affected by the "Direction" factor. The effect of the "Planes" factor (transverse, radial, tangential) on the *Ra* values with regard to the vertical measurements was very low (1.1%), while for the parallel measurements, it was 18.5%.

Table 2. Analysis of variance output—Univariate tests of *Ra* (dependent variable).

Surface Planes		Sum of Squares	df	Mean Square	F	Sig.	Partial Eta Squared
Transverse	Contrast	1383.516	1	1383.516	395.408	<0.001	0.209
	Error	5227.442	1494	3.499			
Radial	Contrast	4972.470	1	4972.470	1421.129	<0.001	0.488
	Error	5227.442	1494	3.499			
Tangential	Contrast	5373.363	1	5373.363	1535.704	<0.001	0.507
	Error	5227.442	1494	3.499			

df: degrees of freedom, F: F-value is the ratio of between-group and within-group variation, Sig.: significance.

4. Conclusions

The results of the current study revealed that the three roughness indexes, *Ra*, *Rz*, and *Rq*, that were studied had a parallel progression with one another. When the measurement was carried out vertically to the wood grain on any of the disks' surfaces (transverse, radial, or tangential), the resulting roughness was noticeably higher compared to when the measurements were implemented in parallel to the wood grain. Although sapwood seems to more frequently exhibit higher surface roughness than the comparable heartwood areas, no discernible differences in roughness between the two types of wood were found. When the roughness measurements were conducted vertically to the wood-grain, the tangential surfaces demonstrated slightly higher roughness values among the three surfaces (transverse, radial, and tangential), with the transverse surfaces showing slightly lower roughness values, and the radial surfaces showing the lowest values. Nevertheless, when the measurements were in parallel to the wood grain, the transverse surfaces had significantly higher roughness values compared to the tangential and radial surfaces. There were no significant variations in surface roughness among the different disks or trunk heights. Conversely, notable variations in roughness were seen amongst the various trees, with Tree No. 3 exhibiting the lowest roughness (and the most significant statistical difference among the trees studied), when compared to the other trees. But, as it happened, each of the studied chestnut trees was found to fit the recognized and described within-tree trends in detail.

The findings of this research are expected to contribute to the rational and thorough utilization of chestnut wood, which constitutes a particularly significant, commercially valuable species of hardwood, as well as to the integration and advancement of fundamental scientific knowledge, which is crucial for guaranteeing the generation of wooden structures characterized by superior surface quality and a longer service life.

Author Contributions: Conceptualization, V.K. and M.C.; methodology, V.K. and M.C.; validation, V.K. and M.C.; formal analysis, M.C. and V.K.; investigation, V.K. and M.C.; resources, M.C.; data curation, V.K.; writing—original draft preparation, V.K. and M.C.; writing—review and editing, V.K. and M.C. All authors have read and agreed to the published version of the manuscript.

Funding: This research received no external funding.

Institutional Review Board Statement: Not applicable.

Informed Consent Statement: Not applicable.

Data Availability Statement: Available upon request to the corresponding author.

Conflicts of Interest: The authors declare no conflicts of interest.

References

1. Kilic, M.; Hiziroglu, S.; Burdurlu, E. Effect of machining on surface roughness of wood. *Build. Environ.* **2006**, *41*, 1074–1078. [CrossRef]
2. Coelho, C.L.; Carvalho, L.M.; Martins, J.M.; Costa, C.A.; Masson, D.; Méausoone, P.J. Method for evaluating the influence of wood machining conditions on the objective characterization and subjective perception of a finished surface. *Wood Sci. Technol.* **2008**, *42*, 181–195. [CrossRef]
3. Söğütlü, C. Determination of the effect of surface roughness on the bonding strength of wooden materials. *BioRes.* **2017**, *12*, 1417–1429. [CrossRef]
4. Magoss, E. General regularities of wood surface roughness. *Acta Silv. Lignaria Hung.* **2008**, *4*, 81–93. [CrossRef]
5. Li, G.; Wu, Q.; He, Y.; Liu, Z. Surface roughness of thin wood veneers sliced from laminated green wood lumber. *Maderas. Cienc. Y Tecnol.* **2018**, *20*, 3–10. [CrossRef]
6. Vassiliou, V.; Barboutis, I.; Kamperidou, V. Strength of corner and middle joints of upholstered furniture frames constructed with black locust and beech wood. *Wood Res.* **2016**, *61*, 495–504.
7. Kamperidou, V.; Aidinidis, E.; Barboutis, I. Impact of structural defects on the surface quality of hardwood species sliced veneers. *Appl. Sci.* **2020**, *10*, 6265. [CrossRef]
8. Tsoumis, G. *Wood Science and Technology, Wood Structure and Properties*; Aristotle University of Thessaloniki, AUTh: Thessaloniki, Greece, 2002; p. 346.
9. Filippou, I. *Chemistry and Chemical Technology of Wood*; Giahoudi-Giapouli Publications: Thessaloniki, Greece, 2014; p. 357.

10. Csanády, E.; Magoss, E.; Tolvaj, L. Surface roughness of wood. In *Quality of Machined Wood Surfaces*; Springer: Berlin, Germany, 2015; Volume 257, pp. 83–236. [CrossRef]
11. Bao, M.; Huang, X.; Zhang, Y.; Yu, W.; Yu, Y. Effect of density on the hygroscopicity and surface characteristics of hybrid poplar compreg. *J. Wood Sci.* **2016**, *62*, 441–451. [CrossRef]
12. Sadoh, T.; Nakato, K. Surface properties of wood in physical and sensory aspects. *Wood Sci. Technol.* **1987**, *21*, 111–120. [CrossRef]
13. Örs, Y.; Gürleyen, L. Effect of the cutting direction, number of cutter and cutter to surface smoothness on wood material for planing, faculty of technical education Gazi Üniversity. *J. Polytech.* **2002**, *5*, 335–339.
14. Malkoçoğlu, A. Machining properties and surface roughness of various wood species planed in different conditions. *Build. Environ.* **2007**, *42*, 2562–2567. [CrossRef]
15. Hecker, M.; Becker, G. Die oberflachen-raunhigkeit mikrotomüberschnittenen Douglasienholzes. *Holz Als Roh-Und Werkst.* **1995**, *53*, 220. [CrossRef]
16. Williams, R.S.; Jourdain, C.; Daisey, G.I.; Springate, R.W. Wood properties affecting finish service life. *J. Coat. Technol.* **2000**, *72*, 35–42. [CrossRef]
17. Mitchell, P.H.; Lemaster, R.L. Investigation of machine parameters on the surface quality in routing soft maple. (Solid Wood Products). *For. Prod. J.* **2002**, *52*, 85–91.
18. Gurau, L.; Mansfield-Williams, H.; Irle, M. Processing roughness of sanded wood surfaces. *Eur. J. Wood Wood Prod.* **2005**, *63*, 43–52. [CrossRef]
19. Lavery, D.J.; Mc Larnon, D.; Taylor, J.M.; Moloney, S.; Atanackovic, S. Parameters affecting the surface finish of planed Sitka spruce. *For. Prod. J.* **1995**, *45*, 445–450.
20. Kamperidou, V.; Aidinidis, E.; Barboutis, I. Surface Roughness of Sliced Veneers in Terms of Defects and Wood Structure Variability–Impact of Mild Hydrothermal Treatment. *Wood Ind./Drv. Ind.* **2022**, *73*, 77–191. [CrossRef]
21. Tan, P.L.; Sharif, S.; Sudin, I. Roughness models for sanded wood surfaces. *Wood Sci. Technol.* **2012**, *46*, 129–142. [CrossRef]
22. Aslan, S.; Coşkun, H.; Kılıç, M. The effect of the cutting direction, number of blades and grain size of the abrasives on surface roughness of Taurus cedar (*Cedrus libani* A. Rich.) woods. *Build. Environ.* **2008**, *43*, 696–701. [CrossRef]
23. Sulaiman, O.; Hashim, R.; Subari, K.; Liang, C.K. Effect of sanding on surface roughness of rubberwood. *J. Mater. Process. Technol.* **2009**, *209*, 3949–3955. [CrossRef]
24. Conedera, M.; Manetti, M.C.; Giudici, F.; Amorini, E. Distribution and economic potential of the sweet chestnut (*Castanea sativa* Mill.) in Europe. *Ecol Mediterr* **2004**, *30*, 179–193. [CrossRef]
25. Voulgaridis, H. Quality and Uses of Wood. Kallipos Repository. 2015. Available online: https://repository.kallipos.gr/handle/11 419/5260 (accessed on 1 February 2024).
26. Vassiliou, V.; Passialis, C.; Voulgaridis, H. Microscopic identification of the wood of Greek forest trees with the help of computers. *Sci. Yaerbook Sch. For. Nat. Environ.* **1993**, *ΙΣΤ*, 331–375.
27. Chavenetidou, M. Anatomical Characteristics and Technical Properties of Chestnut (*Castanea sativa* Mill.) Wood from Coppice Forests in Relation to Its Utilization. Ph.D. Thesis, Aristotle University of Thessaloniki (AUTh), Thessaloniki, Greece, 2009.
28. Adamopoulos, S.; Chavenetidou, M.; Passialis, C.; Voulgaridis, E. Effect of cambium age and ring width on density and fibre length of black locust and chestnut wood. *Wood Res.* **2010**, *55*, 25–36.
29. Chavenetidou, M.; Kakavas, K.V.; Birbilis, D. Shrinkage and Swelling of Greek Chestnut Wood (*Castanea sativa* Mill.) in Relation to Extractives Presence. In *IOP Conference Series: Materials Science and Engineering*; IOP Publishing: Bristol, UK, 2020; Volume 908, p. 012004.
30. Sutcu, A.; Karagoz, U. The influence of process parameters on the surface roughness in aesthetic machining of wooden edge-glued panels (EGPs). *Bioresources* **2013**, *8*, 5435–5448. [CrossRef]
31. *ISO 13061-1*; Physical and Mechanical Properties of Wood—Test Methods for Small Clear Wood Specimens. Part 1: Determination of Moisture Content for Physical and Mechanical Tests. ISO: Geneva, Switzerland, 2014.
32. Magoss, E.; Rozs, R.; Tatai, S. Evaluation of wood surface roughness by confocal microscopy. *Wood Res.* **2022**, *67*, 919–928. [CrossRef]
33. *ISO 21920-2:2021*; Geometrical Product Specifications (GPS)—Surface Texture: Profile. Part 2: Terms, Definitions and Surface Texture Parameters. ISO: Geneva, Switzerland, 2021.
34. Budakci, M.; Cemil Ilce, A.; Gurleyen, T.; Uter, M. Determination of the surface roughness of Heattreated wood materials planed by the cutters of the horizontal milling machine. *BioResources* **2013**, *8*, 3189–3199. [CrossRef]
35. Kamperidou, V.; Barboutis, I. Mechanical strength and surface roughness of thermally modified poplar wood. *PRO Ligno* **2017**, *13*, 107–114.
36. Korkut, D.S.; Korkut, S.; Bekar, I.; Budacki, M.; Dilik, T.; Cakicier, N. The effects of heat treatment on the physical properties and surface roughness of turkish hazel (*Corylous colurna* L.) wood. *Int. J. Mol. Sci.* **2008**, *9*, 1772–1783. [CrossRef] [PubMed]
37. Sandak, J.; Negri, M. Wood surface roughness—What is it? In Proceedings of the 17th International Wood Machining Seminar, Rosenheim, Germany, 26–28 September 2005; Volume 1, pp. 242–250, (PDF) On-Line Measurement of Wood Surface Smoothness. Available online: https://www.researchgate.net/publication/342170861_OnLine_Measurement_of_Wood_Surface_Smoothness (accessed on 1 August 2020).

38. Laina, R.; Sanz-Lobera, A.; Villasante, A.; López-Espí, P.; Martínez-Rojas, J.A.; Alpuente, J.; Sánchez-Montero, R.; Vignote, S. Effect of the anatomical structure, wood properties and machining conditions on surface roughness of wood. *Maderas. Cienc. Y Tecnol.* **2017**, *19*, 203–212. [CrossRef]
39. Kakavas, K.; Chavenetidou, M.; Birbilis, D. Effect of ring shakes on mechanical properties of chestnut wood from a Greek coppice forest. *For. Chron.* **2018**, *94*, 61–67. [CrossRef]
40. Fortino, S.; Metsäjoki, J.; Ronkainen, H.; Bjurhager, I.; Heiemann, S.; Salminen, L.I. Scratch resistance of PEG impregnated green wood: A method for evaluation of swollen wood properties. *Wood Sci. Technol.* **2020**, *54*, 715–735. [CrossRef]
41. Paridah, M.T.; Alia-Syahirah, Y.; Hamdan, H.; Anwar, U.M.K.; Nordahlia, A.S.; Lee, S.H. Effects of anatomical characteristics and wood density on surface roughness and their relation to surface wettability of hardwood. *J. Trop. For. Sci.* **2019**, *31*, 269–277. [CrossRef]

Disclaimer/Publisher's Note: The statements, opinions and data contained in all publications are solely those of the individual author(s) and contributor(s) and not of MDPI and/or the editor(s). MDPI and/or the editor(s) disclaim responsibility for any injury to people or property resulting from any ideas, methods, instructions or products referred to in the content.

Article

Effect of Wood Densification and GFRP Reinforcement on the Embedment Strength of Poplar CLT

Akbar Rostampour-Haftkhani [1,*], Farshid Abdoli [2], Mohammad Arabi [3], Vahid Nasir [4,*] and Maria Rashidi [2,*]

1 Wood Science and Technology, Department of Natural Resources, Faculty of Agriculture and Natural Resources, University of Mohaghegh Ardabili, Ardabil 56199-11367, Iran
2 Centre for Infrastructure Engineering (CIE), School of Engineering, Design and Built Environment, Western Sydney University, Sydney 2145, Australia; abdolifarshid@gmail.com
3 Wood Science and Technology, Department of Natural Resources, Faculty of Agriculture and Natural Resources, University of Zabol, Zabol 11936-53471, Iran; marabi@uoz.ac.ir
4 Department of Wood Science and Engineering, Oregon State University, Corvallis, OR 97331, USA
* Correspondence: arostampour@uma.ac.ir (A.R.-H.); vahid.nasir@oregonstate.edu (V.N.); m.rashidi@westernsydney.edu.au (M.R.)

Citation: Rostampour-Haftkhani, A.; Abdoli, F.; Arabi, M.; Nasir, V.; Rashidi, M. Effect of Wood Densification and GFRP Reinforcement on the Embedment Strength of Poplar CLT. *Appl. Sci.* 2023, *13*, 12249. https://doi.org/10.3390/app132212249

Academic Editors: Alena Očkajová, Martin Kučerka and Richard Kminiak

Received: 9 October 2023
Revised: 9 November 2023
Accepted: 10 November 2023
Published: 12 November 2023

Copyright: © 2023 by the authors. Licensee MDPI, Basel, Switzerland. This article is an open access article distributed under the terms and conditions of the Creative Commons Attribution (CC BY) license (https://creativecommons.org/licenses/by/4.0/).

Abstract: Embedment strength is an important factor in the design and performance of connections in timber structures. This study assesses the embedment strength of lag screws in three-ply cross-laminated timber (CLT) composed of densified poplar wood with densification ratios of 25% and 50%, under both longitudinal (L) and transverse (T) loading conditions. The embedment strength was thereafter compared with that of CLT reinforced with glass-fiber-reinforced polymer (GFRP). The experimental data was compared with results obtained using different models for calculating embedment strength. The findings indicated that the embedment strength of CLT specimens made of densified wood and GFRP was significantly greater than that of control specimens. CLT samples loaded in the L direction showed higher embedment strength compared to those in the T direction. In addition, 50% densification had the best performance, followed by 25% densification and GFRP reinforcement. Modelling using the NDS formula yielded the highest accuracy (mean absolute percentage error = 10.31%), followed by the Ubel and Blub (MAPE = 21%), Kennedy (MAPE = 28.86%), CSA (MAPE = 32.68%), and Dong (MAPE = 40.07%) equations. Overall, densification can be considered as an alternative to GFRP reinforcement in order to increase the embedment strength in CLT.

Keywords: embedment strength; densification; cross-laminated timber

1. Introduction

Cross-laminated timber (CLT) is a type of wood product consisting of at least three orthogonal layers of solid-sawn lumber bonded with adhesive, fasteners, or wooden dowels [1]. Timber construction with CLT has gained popularity in both residential and commercial applications.

Due to the load-bearing resistance of CLT, the efficacy of wood structures is highly dependent on the applied connections [2,3]. In CLT structural systems, connections are essential for lateral force transfer and energy dissipation [4]. Carpenter-made mortise and tenon joints are typically found in traditional wood structures, but these conventional connection types are rarely used in contemporary wood structures. Instead, a variety of standard metal connectors and dowels are utilized [5–7]. Various factors can impact the load-bearing capacity of connections between cross-laminated timber (CLT) elements, including the type of wood species and technology used in CLT production, the specific design of the engineering connection, environmental conditions such as ambient temperature and air humidity, the type of applied load (static or dynamic), and the quality of workmanship [8,9].

Connections and fasteners in timber structures are subjected to numerous external loading conditions, including the lateral load. Estimating the embedment strength of CLT is essential, especially under lateral load conditions. In general, embedment strength is a system property representing the resistance of wood to laterally loaded embedment fasteners. When fasteners are inserted into the side face or the narrow face of cross-laminated timber (CLT) at different angles relative to the fastener axis and load, the resulting system property encompasses a range of inputs from the layers and laminations, which is due to the varied load-grain angles [7].

Due to considerable variations in embedment strength, deformation, ductility, and failure modes between the longitudinal and transverse layers, the orthogonal lamination complicates the connection properties of CLT [1]. Additionally, due to the orthogonal lamination, the embedment locations of dowels in CLT become more complicated. It is possible to locate dowels parallel and perpendicular to the grain within and between a single lamina in one layer as well as between layers when dowels are installed on the narrow face of CLT [1]. Similarly, when installed on the flat side of the CLT panel, the dowels may penetrate into multiple layers. Therefore, it is necessary to evaluate the embedment strength in wood structures. The embedment strength of wood is necessary for estimating the capacity of structural timber connectors using dowel-type fasteners. This strength is derived from expressions that are based on tests carried out on timber. Dowel-type connectors are the most frequent form of joint used in modern construction, and they are typically made of wood or metal [10]. Dowel-type connections provide a number of benefits, one of which is their ductile characteristic, which makes it possible for considerable relative deformations and rotations to occur between the wood pieces [11]. Therefore, the embedding strength is not a specific attribute of the material but rather a property of the system. Different empirical equations have been proposed to estimate embedment strength. However, a thorough examination is necessary to develop novel approaches that include relevant derived factors for the designer's benefit [10].

Numerous factors influence the embedment strength of wood and engineered wood products (EWPs), including the type of wood species, density, moisture content, loading direction, dowel diameter, and so on [12].

Multiple studies have evaluated the embedment strength of connectors in CLT. Blaß and Uibel [13] and Uibel and Blaß [14–16] conducted the first investigations of the embedment strength of laterally loaded dowel-type fasteners in CLT. Santos et al. [17] observed that wood density and embedment strength are related. This positive correlation was also reported by [16]. A study on the nail-bearing strength of hybrid CLT composed of Japanese larch and yellow poplar [18] revealed that increasing the higher ratio of minor lamina thickness to nail depth resulted in a lower embedment strength in the hybrid CLT. To obtain effective bearing resistance of the nail connection, the length of the nail used for the mixed CLT should be chosen based on the thickness of the minor lamina. Several additional studies [19–22] have investigated the embedment strength of connectors in CLT and wood-based products. They worked on the embedment strength of EWPs such as CLT made of various wood species.

The timber species used to manufacture CLT significantly affects its properties. Poplar is a fast-growing tree, which is advantageous in nations with a wood supply shortage. Several studies have examined the CLT properties of fast-growing timber species [23–29].

It is essential to reinforce the wood, notably at vulnerable connection points [30]. In addition, the low density of certain wood species, such as poplar, may results in undesirable mechanical properties. There are various techniques to reinforce EWPs and improve the mechanical strength of timber structures, including but not limited to reinforcement using metals or fiber-reinforced polymers (FRPs). According to Saribiyik and Akguuml [31], reinforcing the connections aims to preserve the continuity of structures and lessen the drawbacks of connection elements with nails and bolts. Several investigations on concrete [32,33], metal plates [32,34–36], rods, and bars as reinforcement in wood-based products have been performed.

Additionally, FRPs can be used in different ways for reinforcement [37]. Due to their strength-to-weight ratio, glass-fiber-reinforced polymers (GFRPs) appear to be an optimal fiber type for reinforcing wood components [31]. Douglas fir split timber stringers were strengthened for shear and bending forces with GFRP layers. Depending on the severity of beam fracture prior to reinforcement, the proposed strengthening design increased the stiffness [38]. Hay et al. [39] showed that diagonal (0–90° lay-up) GFRP layers were more effective than vertical ones in shear-strengthening creosote-treated Douglas fir beams with horizontal fractures at their extremities. The mechanical characteristics of connecting points of fiber-reinforced longitudinal notched lap joints fabricated from black pine lumber were investigated [31]. The results demonstrated that GFRP could be utilized as a connecting mechanism for timber members. Wu et al. [40] created a GFRP wood-affixed connection as a replacement due to the potentially corrosive character of steel-plated pine wood connectors. According to their findings, the majority of GFRP wood-bolted connectors failed due to the bearing failure of the bolt openings in the wood panels when subjected to lateral tension load.

Densification is another method for enhancing the mechanical properties of timber and timber-based products that increases the strength, hardness, and abrasion resistance of timber [41]. Densified timber can be utilized by creating laminates and processing the densified wood with a similar construction to that of laminated veneer lumber (LVL), glue-laminated timber (Glulam), and CLT. Feng and Chiang investigated the mechanical strength of CLTs manufactured from densified wood [42]. The use of densified timber enhanced the CLT's bending strength and rigidity, with an increase in modulus of elasticity (MOE) and modulus of rupture (MOR). A similar study on the use of densified wood was conducted by Salca et al. [43], who manufactured and tested densified plywood. MOE, MOR, and shear strength increased, as did the bonding quality of the adhesives between panels.

There is currently a knowledge gap on evaluating the effects of densification and GFRP reinforcement on the embedment strength of CLT manufactured from fast-growing poplar wood. Thus, this study aims to compare the embedment strength of CLT made of densified timber with that of reinforced with GFRP in two loading directions in order to determine which reinforcing technique may be more advantageous. In addition, different approaches are used to calculate the embedment strength and find the model yielding minimum error compared to the experimental data.

2. Materials and Methods

2.1. Materials and CLT Fabrication

Three-ply CLT panels made of air-dried poplar (*Populus alba*) wood with moisture content of around 12% and oven-dry density of 400 ± 10 kg/m^3 were used in this research. CLT panels were produced in four groups. The thickness of the boards was 2 cm. One group was reinforced with GFRP using a bidirectional (0°/90°) E-glass fiber fabric (fiber tensile modulus and density of 70 GPa and 2.55 gr/cm^3, respectively), as shown in Figure 1. Three GFRP layers were added to each surface. Two groups were manufactured with densified wood in two densification ratios (25% and 50%). In order to manufacture densified wood, lumber panels were placed between heated platens in a hydraulic press and compressed. Then, CLT panels were made from the densified wood. The last group consisted of CLT with non-densified wood, considered as a control group (0). A summary of the sample groups is detailed in Table 1. All components were cold-pressed for 150 min and pressure of 1 MPa with one-component polyurethane glue.

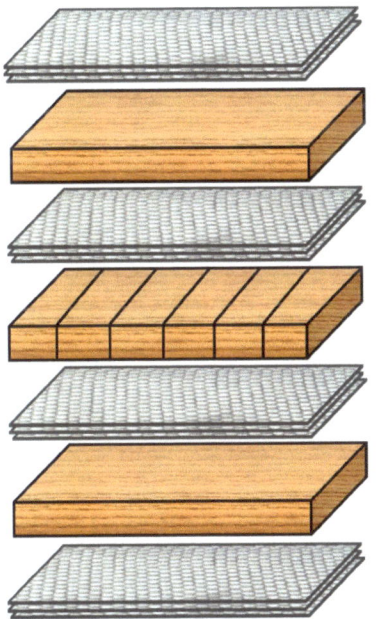
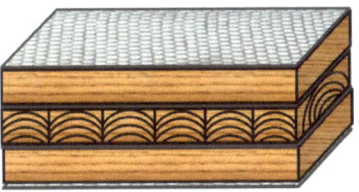

Figure 1. GFRP arrangement in CLT layers.

Table 1. Sample groups and repetitions.

Sample Groups	Repetitions	Description
Control (unreinforced) 0L, 0T	10	No reinforcement (Loading direction: L = longitudinal, T = transverse)
25L	10	Reinforcement with 25% densification (loading direction: longitudinal)
25T	10	Reinforcement with 25% densification (loading direction: transverse)
50L	10	Reinforcement with 50% densification (loading direction: longitudinal)
50T	10	Reinforcement with 50% densification (loading direction: transverse)
GFRPL	10	Reinforcement with GFRP (loading direction: longitudinal)
GFRPT	10	Reinforcement with GFRP (loading direction: transverse)

Embedment Strength Modelling and Experiments

After preparing the CLT panel, specimens with dimensions of 15 cm × 8 × cm × 6 cm (length × width × thickness) were cut from the panel for the embedment strength test (Figure 2). A lag screw (s) with a diameter of 8 mm was used as a fastener. The characteristics of the fastener are detailed in Table 2.

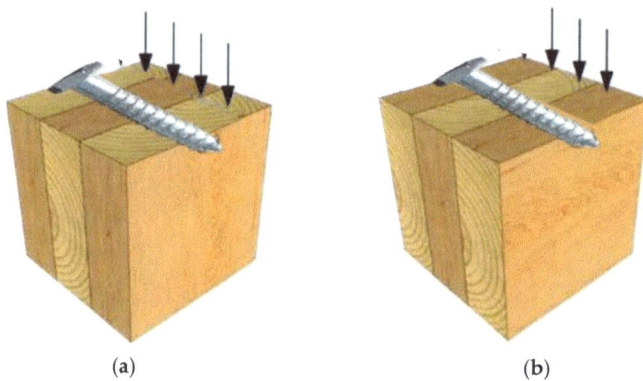

Figure 2. Loading directions of CLT samples for embedment strength: (**a**) loading in longitudinal (L) direction; (**b**) loading in transverse (T) direction.

Table 2. Characteristics of lag screw.

Lag Screw (DIN 571)	Length of the Screw (L) (mm)	Head Height (K) (mm)	Da (mm)	Ds (mm)	Length of the Thread (b) (mm)
	54	5.5	8.5	7	44

Finally, the embedment strength (σ) was calculated using $\sigma\left(\frac{N}{mm^2}\right) = \frac{F}{A}$, in which F is the yield load obtained from the 5% offset method (N) and A is the embedment area (mm^2).

To determine the embedment strength, specimens were subjected to loading in two directions of longitudinal (L) and transverse (T), as shown in Figure 3 using a Hounsfield testing equipment model 0308 (Figure 3) and a loading pace of 5 mm/min.

Figure 3. The loading apparatus of CLT samples for embedment strength testing.

2.2. Statistical Analysis

The data was assessed statistically as a complete factorial design consisting of two factors using the SPSS 25 program. The densification ratio (in three levels: 0, 25, and 50) and two CLT loading directions (L and T directions) were the two factors that were taken into consideration for the statistical analysis. ANOVA test used to analyze those factors statistically through the main and interaction effects. Afterward, the results (means) of densified CLT samples were compared to those reinforced with GFRP. Four different sample groups, each with ten replicates, were put through their paces throughout the testing process. These sample groups included control (unreinforced) CLT specimens, CLT samples with 25% densified layers, CLT samples with 50% densified layers, and CLT samples reinforced with GFRP, all of which were loaded in both the longitudinal (L) and transverse (T) directions. The statistical differences between the means were analyzed using the multiple range test developed by Duncan with a confidence level of 95%.

2.3. Embedment Strength Equations

Currently, a number of distinct models of computation for CLT embedment strength are available. The most important calculation models are discussed in this part so that a comparison can be made between them.

The Kennedy [44], NDS [45], Ubel and Blub [46], Dong [2], and CSA [47] equations were used in this study to predict the embedment strength of CLT. These equations were developed based on various factors, such as density, loading direction, and so on.

Kennedy et al. conducted a comprehensive study including about 720 embedment tests on Canadian cross-laminated timber (CLT) using lag screws, as well as 360 tests using self-drilling screws. The experiments included a range of screw sizes from 6.0 mm to 19.1 mm. A regression model that is not influenced by the panel layup and fastener diameter was constructed. According to Kennedy's model, the embedment strength is calculated as follows:

$$f_{\theta,avg,Ken} = \frac{80(\rho_{12} - 0.12)^{1.11}}{1.07(\rho_{12} - 0.12)^{-0.07} sin^2\theta + cos^2\theta} \quad (1)$$

where $f_{\theta,avg}$ is average embedment strength (MPa), ρ_{12} is density at 12% moisture content (g/cm^3), and θ is loading angle relative to the grain of face layer (°).

Based on the NDS model, the strength of the face layer's embedment is linked to the "effective" bearing length of the fastener. This bearing length is changed proportionally based on the embedment strengths of both the transverse layer and the longitudinal layer.

According to the NDS model, the embedment strength is calculated as follows:

$$f_{\theta,avg,NDS} = \frac{l_\parallel}{l_\rho} \cdot \frac{77 G_0}{0.36 G_0^{-0.45} d^{0.5} sin^2\theta + cos^2\theta} + \frac{l_\perp}{l_\rho} \cdot \frac{77 G_0}{0.36 G_0^{-0.45} d^{0.5} cos^2\theta + sin^2\theta} \quad (2)$$

where G_0 is measured relative density for the species or species group based on oven-dry mass and volume; $l_\parallel$ is the lag screw embedment length in parallel layer (s); $l_\perp$ is the lag screw embedment length in cross layer (s); l_ρ is the embedment length of the lag screw in CLT specimen; and d is fastener diameter (mm).

Ubel and Blub conducted pioneering research on the lateral loading behavior of dowel-type fasteners in cross-laminated timber (CLT) panels. The researchers conducted experiments on smooth dowels ranging in diameter from 8 mm to 24 mm. These dowels were tested in both three- and five-layer cross-laminated timber (CLT) components. The dowels were strategically placed at gaps of one to three layers and subjected to loading at angles of 0°, 45°, and 90° with respect to the grain orientation of the face layer. The model is formulated as a mathematical expression that incorporates the variables of fastener diameter, overall wood density of the panel, and loading direction in relation to the strength axis of the panel, namely the grain direction of the surface layers of the cross-laminated timber (CLT) panel.

According to Ubel and Blub's model, the embedment strength is calculated as follows:

$$f_{\theta,avg,UB} = 111.7(1 - 0.016d)\, \rho_{12}^{1.16} \times \left[\frac{\sum_{i=1}^{n} t_{0,i}}{t\left(1.2\sin^2\theta + \cos^2\theta\right)} + \frac{\sum_{j=1}^{n-1} t_{90,j}}{t\left(1.2\cos^2\theta + \sin^2\theta\right)} \right] \quad (3)$$

where $t_{0,i}$ is the thickness of the CLT longitudinal layer i, and $t_{90,j}$ the thickness of the CLT transverse layer i.

According to Dong's model, the embedment strength is calculated as follows:

$$f_{\theta,avg,Dong} = 336.4(0.45 - 0.02d)\, \rho_{12} \times \left(\frac{R_t}{1.41\cos^2\theta + \sin^2\theta} + \frac{1 - R_t}{1.41\sin^2\theta + \cos^2\theta} \right) \quad (4)$$

where R_t the ratio between the total thickness of the transverse layers and the CLT thickness.

According to the CSA's model, the embedment strength is calculated as follows:

$$f_{\theta,avg,CSA} = \frac{0.9 \times 82_{\rho_{12}}(1 - 0.01d)}{0.9 \times 2.27\sin^2\theta + \cos^2\theta} \quad (5)$$

All of the notation used in all equations is mentioned above. Mean absolute percentage error (MAPE) values were used to evaluate the prediction performance of each model as follows (Equation (6)):

$$MAPE = \frac{1}{n} \sum_{i=1}^{n} \left(\frac{|y_i - y_p|}{y_i} \right) 100 \quad (6)$$

where y_i is the experimental value, y_p is the predicted value, and n is the total amount of data. The lower the MAPE values, the smaller the difference between experimental and predicted values.

3. Results and Discussion

Effect of Densification Ratio and Loading Direction on Embedment Strength

With a confidence level of 95% (p-value > 0.05), the results of the analysis of variance (ANOVA) shown in Table 3 suggest that both the densification ratio and the loading direction had significant main impacts on embedment strength.

Table 3. ANOVA table for main and interaction effects of densification ratio and loading direction on embedment strength of CLT samples.

Property	Source	df	Mean Square	F	Sig.
Embedment strength	Densification ratio	2	2079.221	124.640	0.000 **
	Loading direction	1	151.877	9.104	0.004 **
	Densification ratio × loading direction	2	4.520	0.271	0.764 ns

** significant at 99% confidence level, ns non-significant.

As can be seen in Table 4, when compared to the control specimens, the embedment strength of CLT samples improved by 46% and 66.8%, respectively, with a 25% and 50% increase in densification ratio. Furthermore, it was discovered that the embedment strength of the CLT samples was 8% greater in the longitudinal (L) direction compared to the transverse (T) direction when the main effect of the loading direction of the CLT samples on embedment strength was considered. This was due to the fact that both of these factors were changed simultaneously. In addition, the findings showed that the embedment strength of CLT samples with a densification ratio of 0 (control samples), 25%, and 50% in the longitudinal loading direction of CLT samples was 10.9%, 5.4%, and 8.8% greater, respectively, compared to the transverse loading direction.

Table 4. Main and interaction effects of densification ratio and loading direction on embedment strength of CLT samples with Duncan test results.

	Main Effects		
Densification ratio	0	29.8 (4.65) *	A **
	25	43.7 (3.4)	B
	50	49.7 (4.76)	C
Loading direction	T	39.44 (9.29)	A
	L	42.62 (9.41)	B
Interaction effects			
Densification ratio	Loading direction	Embedment strength (MPa)	
0	T	28.24 (4.88)	A
	L	31.33 (4.07)	A
25	T	42.52 (3.61)	B
	L	44.8 (2.92)	BC
50	T	47.58 (4.19)	C
	L	51.75 (4.54)	D

* The values in parentheses represent standard deviation. ** The letters show Duncan test results.

Comparing the embedment strength of CLT samples manufactured from densified wood and those reinforced with GFRP was one of the goals of this investigation. The embedment strength of the GFRP-reinforced CLT samples was compared with that of CLT samples constructed out of densified wood, and the average embedment strength of the CLT specimens is illustrated in Figure 4. The highest embedment strength belonged to the 50L samples (51.75 MPa), while the lowest embedment strength belonged to the 0T samples (28.24 MPa). The embedment strength of the CLT samples in the L direction was greater than in the T direction. More specifically, the embedment strength of 0L, 25L, 50L, and GFRPL samples was 10.9%, 5.4%, 8.8%, and 6.8% more, respectively, than that of their counterpart samples tested in transverse direction. Reinforcement with 25% densification, 50% densification, and GFRP improved the embedment strength by 43%, 65.2%, and 43.4%, respectively, compared to the unreinforced samples in the L direction (0L). On the other hand, reinforcement with 25% densification, 50% densification, and GFRP improved the embedment strength by 50.6%, 68.5%, and 49%, respectively, compared to the unreinforced samples in the T direction (0T). In addition, no significant difference was observed in the embedment strength of reinforced CLT samples with 25% densification and GFRP in both directions (L and T). Therefore, it can be concluded that CLT with the same embedment strength can be produced using wood with a densification ratio of 25% rather than reinforcing it with GFRP. This alternate method demonstrates more efficiency in terms of both its cost and implementation. Previously, it was reported that embedment strength perpendicular to the grain was lower than that parallel to the grain [1,2,48,49]. Reinforcement with densification enhanced the wood density resulting in higher embedment strength of CLT samples. From a designer's point of view, wood density and dowel diameter are the main properties of design. There are also some approaches where density is the only parameter [22,50]. Concerning face side insertion in CLT, Uibel and Blab [15,46] evaluated the load-bearing capacity of dowels placed in manufactured with four distinct arrangements and computed the embedment strength. The CLT embedment strength prediction model was suggested based on the variable CLT lamination characteristics.

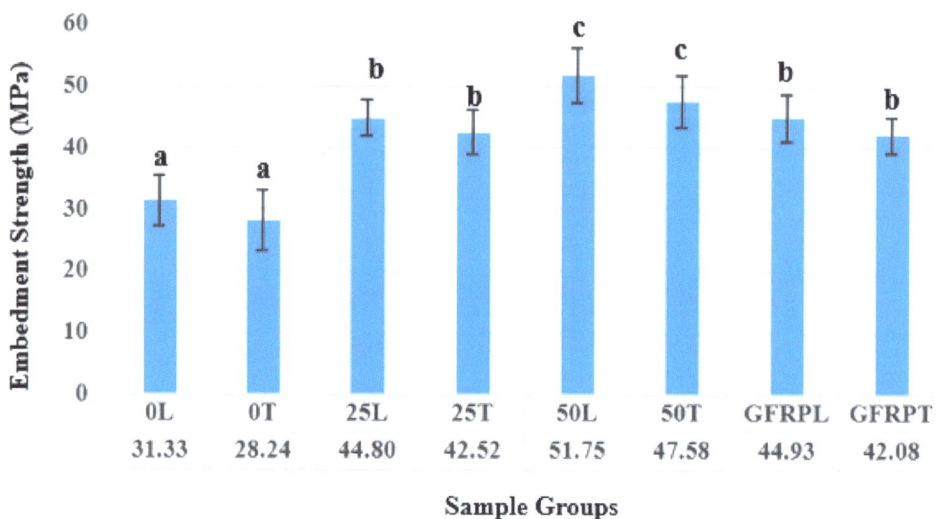

Figure 4. Means and standard deviations of embedment strength of control, densified, and GFRP-reinforced CLT specimens.

The load displacement of control, densified, and GFRP-reinforced CLT specimens is shown are Figure 5. Accordingly, the yield points of the CLT samples in the L direction were more than those in T direction. The yield points of 0L, 25L, 50L, and GFRPL were 8636 (N), 11,900 (N), 16,080 (N), and 13,720 (N), respectively. The highest yield point in the L direction belonged to the CLT sample reinforced with 50% densified wood. On the other hand, the yield points of 0T, 25T, 50T, and GFRPT were 7947 (N), 11,450 (N), 13,152 (N), and 12,120 (N), respectively. The highest yield point in the L direction belonged to the CLT sample reinforced with 50% densified wood.

The failure modes of CLT specimens are depicted in Figure 6. The typical failure modes of tested CLT specimens are different in the situation where the lag screw is loaded in the L and T direction of the CLT specimens. Cracks occurred along the grain when the load was applied along with the L direction, as shown in Figure 6 (0L, 25L, and 50L). Cracks in 25L and 50L show that the densification reduced the length of the cracks compared to the unreinforced samples (0L). However, compression failure occurred in the embedment surface, as shown in Figure 6 (0T, 25T, 50T, GFRPL, and GFRPT), or cracks appeared on both edges of the cross layers, as shown in Figure 6 (0T), when the CLTs were loaded in the T direction. The compression failure might occur layer by layer, which results in stress redistribution [1].

The results of the prediction models in comparison with experimental results are shown in Figure 7. NDS showed the most accurate results compared to the experimental findings, followed by the Ubel and Blub and then the Kennedy and CSA formulas. Finally, the formula of Dong et al. showed the lowest accuracy in terms of the prediction of experimental results. Previously, all of these equations were applied to predict the embedment strength of fasteners in EWPs, such as CLT. Accordingly, various results were obtained regarding the accuracy of these equations. The accuracy of the equations might be related to factors including moisture content, density, diameter of fastener, or the loading direction.

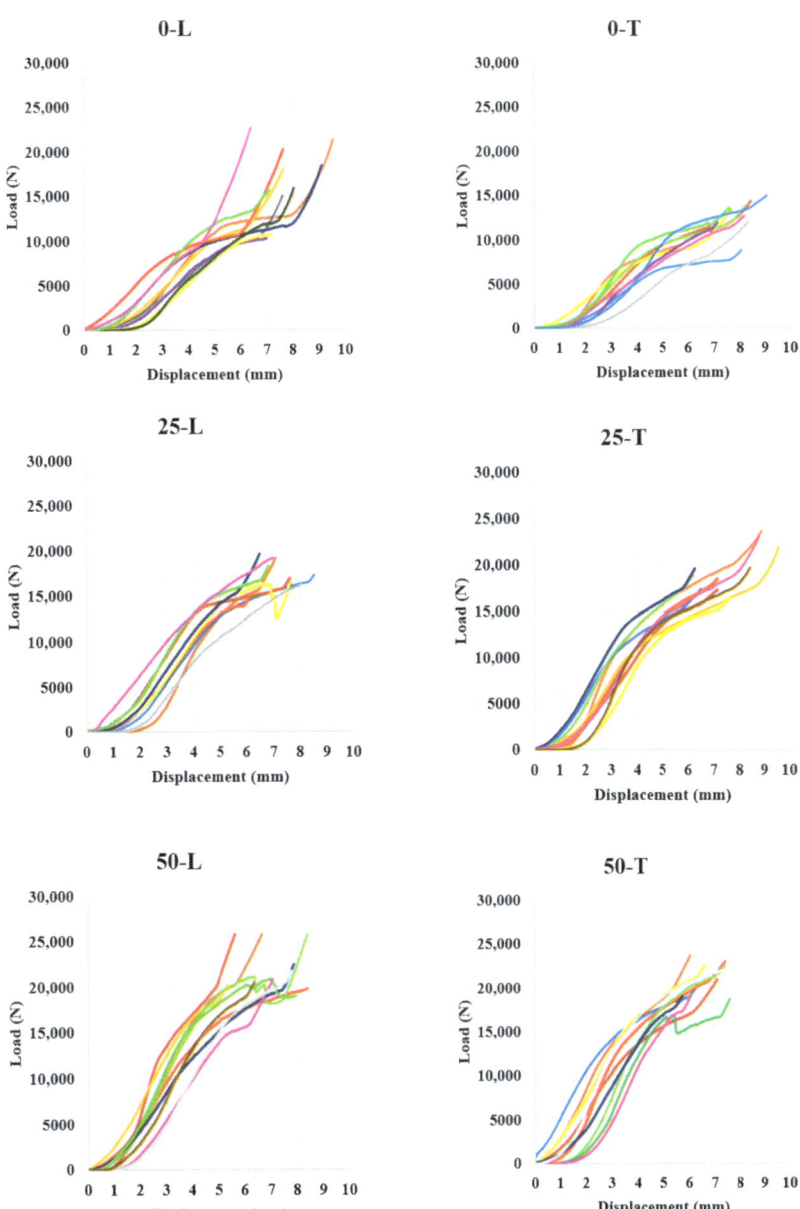

Figure 5. *Cont.*

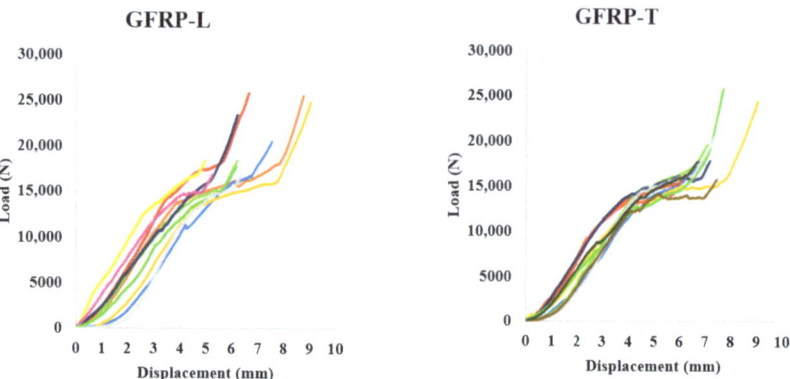

Figure 5. Load-displacement curves of the reinforced and unreinforced CLT samples.

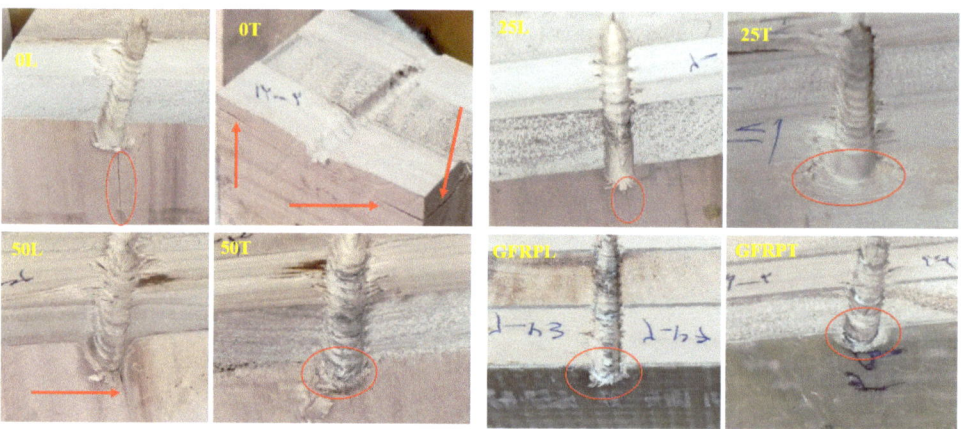

Figure 6. Failure modes of reinforced and unreinforced CLT specimens in L and T loading directions.

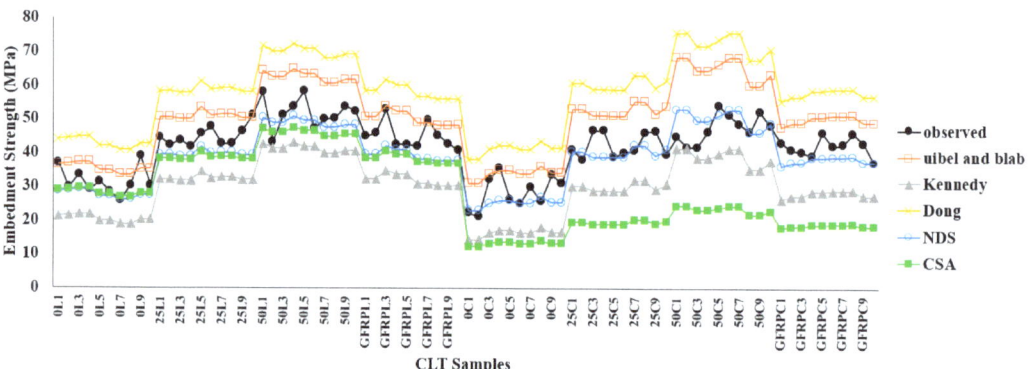

Figure 7. Comparison of the observed and predicted values of the embedment strength.

The MAPE of all models is detailed in Table 5. The NDS formula showed the highest accuracy (10.31%). After that, the Ubel and Blub formula showed the highest accuracy (21.3%). On the other hand, Dong et al.'s formula showed the lowest accuracy (40.07%). These findings indicate that the NDS and Ubel and Blub formulas are reliable in predicting embedment strength in the tests related to loading directions and reinforcement factors. The MAPE classification approach, which was established by Lewis [51], was used in this investigation for the purpose of determining how accurately prediction models performed. This categorization assigns models into one of four accuracy categories based on their mean absolute percentage error (MAPE): excellent (MAPE < 10%), acceptable (MAPE 10–19%), fair (MAPE 20–49%), and weak (MAPE $\geq$ 50%). Future studies could consider developing machine learning modeling for embedment strength prediction. In addition, nondestructive testing methods such as acoustic emission monitoring could be used to further investigate the failure mechanism [52].

Table 5. MAPE of equations for predicting embedment strength of CLT specimens.

MAPE (%)				
Ubel and Blab	Kennedy	Dong et al.	NDS	CSA
21.30	28.86	40.07	10.31	32.68

4. Conclusions

This work evaluated the embedment strength of 3-ply poplar CLT samples reinforced with GFRP and densified wood (25% and 50%) in two loading directions (L and T). Overall, reinforcement improved the embedment strength of CLT samples. The embedment strength of CLT samples in the L loading direction (outer layers to the same fiber direction) was greater than those in the T loading direction. CLT samples made from 50% densified poplar wood showed the highest embedment strength (50L: 51.75 MPa and 50T: 47.58 MPa). However, there was no significant difference in embedment strength between CLT samples reinforced with GFRP and CLT samples made from 25% densified wood (25L: 44.8 MPa, GFRPL: 44.93 MPa, 25T: 42.52 MPa, and GFRPT: 42.08 MPa). Modelling using the NDS formula yielded the highest accuracy (MAPE = 10.31%), followed by the Ubel and Blub (MAPE = 21%), Kennedy (MAPE = 28.86%), CSA (MAPE = 32.68%), and Dong (MAPE = 40.07%) equations. Densification can be considered as an alternative to GFRP reinforcement in order to increase the embedment strength in CLT. Statistical analysis showed that in terms of main effects, densification ratio and loading direction significantly affect the embedment strength. However, in terms of interaction effects, both densification ratio and loading direction at the same time had no significant effect on the embedment strength.

Further studies are recommended to evaluate the embedment strength of CLT reinforced with GFRP and densified wood made of more than three layers. Further studies are recommended to evaluate the embedment strength of various types of nails and screws in CLT reinforced with GFRP and densified wood. According to the results of this study, considering the cost of the products, 25% densification might be used instead of GFRP for CLT reinforcement. However, further investigations regarding the effects of densification on CLT properties, such as bending, glue line, and debonding, are required.

Author Contributions: Conceptualization, A.R.-H. and M.A.; methodology, A.R.-H. and M.A.; formal analysis, A.R.-H., F.A. and M.A.; resources, A.R.-H., F.A. and M.A.; writing—original draft preparation, F.A.; writing—review and editing, A.R.-H., V.N. and F.A.; visualization, F.A.; supervision, A.R.-H. and M.R.; project administration, A.R.-H., V.N. and M.R. All authors have read and agreed to the published version of the manuscript.

Funding: This research received no external funding.

Institutional Review Board Statement: Not applicable.

Informed Consent Statement: Not applicable.

Data Availability Statement: The data is not publicly available as it is part of some ongoing research by the authors. However, they may be available upon reasonable request from corresponding authors.

Conflicts of Interest: The authors declare no conflict of interest.

References

1. Dong, W.; Li, Q.; Wang, Z.; Zhang, H.; Lu, X.; Gong, M. Effects of embedment side and loading direction on embedment strength of cross-laminated timber for smooth dowels. *Eur. J. Wood Wood Prod.* **2020**, *78*, 17–25. [CrossRef]
2. Dong, W.; Wang, Z.; Zhou, J.; Zhang, H.; Yao, Y.; Zheng, W.; Gong, M.; Shi, X. Embedment strength of smooth dowel-type fasteners in cross-laminated timber. *Constr. Build. Mater.* **2020**, *233*, 117243. [CrossRef]
3. Rashidi, M.; Hoshyar, A.N.; Smith, L.; Samali, B.; Siddique, R. A comprehensive taxonomy for structure and material deficiencies, preventions and remedies of timber bridges. *J. Build. Eng.* **2021**, *34*, 101624. [CrossRef]
4. Bora, S.; Sinha, A.; Barbosa, A.R. Effect of Short-Term Simulated Rain Exposure on the Performance of Cross-Laminated Timber Angle Bracket Connections. *J. Archit. Eng.* **2022**, *28*, 04022025. [CrossRef]
5. Ling, T.; Mohrmann, S.; Li, H.; Bao, N.; Gaff, M.; Lorenzo, R. Review on research progress of metal-plate-connected wood joints. *J. Build. Eng.* **2022**, 105056. [CrossRef]
6. Ringhofer, A.; Brandner, R.; Blaß, H.J. Cross laminated timber (CLT): Design approaches for dowel-type fasteners and connections. *Eng. Struct.* **2018**, *171*, 849–861. [CrossRef]
7. Ling, Z.; Rong, X.; Xiang, Z. Laterally loaded performance of single dowel-type fastener used for steel plate-to-timber connections. In *Structures*; Elsevier: Amsterdam, The Netherlands, 2021; pp. 1985–1997.
8. Sydor, M.; Rogoziński, T.; Stuper-Szablewska, K.M.; Starczewski, K. The accuracy of holes drilled in the side surface of plywood. *BioResources* **2020**, *15*, 117–129. [CrossRef]
9. Sydor, M.; Majka, J.; Rychlik, M.; Turbański, W. Application of 3D Scanning Method to Assess Mounting Holes' Shape Instability of Pinewood. *Materials* **2023**, *16*, 2053. [CrossRef]
10. Yurrita, M.; Cabrero, J.M. New criteria for the determination of the parallel-to-grain embedment strength of wood. *Constr. Build. Mater.* **2018**, *173*, 238–250. [CrossRef]
11. Jorissen, A.; Fragiacomo, M. General notes on ductility in timber structures. *Eng. Struct.* **2011**, *33*, 2987–2997. [CrossRef]
12. Zhang, X.; Yin, H.; Zhao, E.; Li, S.; Liu, Q. Experimental investigation on embedment strength of bamboo-based composite prepared with the inorganic adhesive. *J. Build. Eng.* **2023**, *76*, 107323. [CrossRef]
13. Blaß, H.J.; Uibel, T. *Tragfähigkeit von Stiftförmigen Verbindungsmitteln in Brettsperrholz*; Karlsruher Institut für Technologie: Karlsruhe, Germany, 2007.
14. Uibel, T.; Blaß, H.J. Edge joints with dowel type fasteners in cross laminated timber. In Proceedings of the CIB-W18 Meeting, Bled, Slovenia, 8 August 2007.
15. Uibel, T.; Blaß, H.J. Joints with dowel type fasteners in CLT structures. In Proceedings of the Focus Solid Timber Solutions-European Conference on Cross Laminated Timber (CLT), Bath, UK, 21–22 May 2013; pp. 119–136.
16. HJ, B. Load-carrying capacity of joints with dowel-type fasteners and interlayers. In Proceedings of the CIB-W18, Delft, 2000, Florence, Italy, 16–19 August 2000.
17. Santos, C.; De Jesus, A.; Morais, J.; Lousada, J. A comparison between the EN 383 and ASTM D5764 test methods for dowel-bearing strength assessment of wood: Experimental and numerical investigations. *Strain* **2010**, *46*, 159–174. [CrossRef]
18. Kim, K. Predicting Nail Withdrawal Resistance and Bearing Strength of Cross-laminated Timbers from Mixed Species. *BioResources* **2021**, *16*, 4027–4038. [CrossRef]
19. Ottenhaus, L.-M.; Li, M.; Smith, T. Analytical derivation and experimental verification of overstrength factors of dowel-type timber connections for capacity design. *J. Earthq. Eng.* **2022**, *26*, 2970–2984. [CrossRef]
20. Ou, J.; Yang, X.; Xu, Q. Experimental research on the embedment strength of cross-laminated timber. *Sichuan Build. Sci.* **2020**, *46*, 28–35.
21. Hongyuan, T.; Menglin, M.; Yuan, Y.; Ruizhong, L.; Zhaoyang, Y. Experimental study on dowel bearing strength of cross-laminated-timber (CLT) panels. *J. Build. Struct.* **2022**, *43*, 175.
22. Sandhaas, C.; Ravenshorst, G.; Blass, H.; Van de Kuilen, J. Embedment tests parallel-to-grain and ductility aspects using various wood species. *Eur. J. Wood Wood Prod.* **2013**, *71*, 599–608. [CrossRef]
23. Abdoli, F.; Rashidi, M.; Rostampour-Haftkhani, A.; Layeghi, M.; Ebrahimi, G. Withdrawal performance of nails and screws in cross-laminated timber (CLT) made of Poplar (*Populus alba*) and Fir (*Abies alba*). *Polymers* **2022**, *14*, 3129. [CrossRef]
24. Abdoli, F.; Rashidi, M.; Rostampour-Haftkhani, A.; Layeghi, M.; Ebrahimi, G. Effects of fastener type, end distance, layer arrangement, and panel strength direction on lateral resistance of single shear lap joints in cross-laminated timber (CLT). *Case Stud. Constr. Mater.* **2023**, *18*, e01727. [CrossRef]
25. Rostampour Haftkhani, A.; Rashidi, M.; Abdoli, F.; Gerami, M. The effect of GFRP wrapping on lateral performance of double shear lap joints in cross-laminated timber as a part of timber bridges. *Buildings* **2022**, *12*, 1678. [CrossRef]
26. Kramer, A.; Barbosa, A.R.; Sinha, A. Viability of hybrid poplar in ANSI approved cross-laminated timber applications. *J. Mater. Civ. Eng.* **2014**, *26*, 06014009. [CrossRef]

27. Wang, Z.; Fu, H.; Chui, Y.-H.; Gong, M. Feasibility of using poplar as cross layer to fabricate cross-laminated timber. In Proceedings of the World Conference on Timber Engineering, Quebec City, QC, Canada, 10–14 August 2014.
28. Hematabadi, H.; Madhoushi, M.; Khazaeian, A.; Ebrahimi, G. Structural performance of hybrid Poplar-Beech cross-laminated-timber (CLT). *J. Build. Eng.* **2021**, *44*, 102959. [CrossRef]
29. Hematabadi, H.; Madhoushi, M.; Khazaeyan, A.; Ebrahimi, G.; Hindman, D.; Loferski, J. Bending and shear properties of cross-laminated timber panels made of poplar (*Populus alba*). *Constr. Build. Mater.* **2020**, *265*, 120326. [CrossRef]
30. Lukaszewska, E.; Fragiacomo, M.; Johnsson, H. Laboratory tests and numerical analyses of prefabricated timber-concrete composite floors. *J. Struct. Eng.* **2010**, *136*, 46–55. [CrossRef]
31. Saribiyik, M.; Akgül, T. GFRP bar element to strengthen timber connection systems. *Sci. Res. Essays* **2010**, *5*, 1713–1719.
32. Miotto, J.L.; Dias, A.A. Evaluation of perforated steel plates as connection in glulam–concrete composite structures. *Constr. Build. Mater.* **2012**, *28*, 216–223. [CrossRef]
33. Nie, Y.; Valipour, H. Experimental and numerical study of long-term behaviour of timber-timber composite (TTC) connections. *Constr. Build. Mater.* **2021**, *304*, 124672. [CrossRef]
34. Clouston, P.; Bathon, L.A.; Schreyer, A. Shear and bending performance of a novel wood–concrete composite system. *J. Struct. Eng.* **2005**, *131*, 1404–1412. [CrossRef]
35. Otero-Chans, D.; Estévez-Cimadevila, J.; Suárez-Riestra, F.; Martín-Gutiérrez, E. Experimental analysis of glued-in steel plates used as shear connectors in Timber-Concrete-Composites. *Eng. Struct.* **2018**, *170*, 1–10. [CrossRef]
36. Chybiński, M.; Polus, Ł. Mechanical behaviour of aluminium-timber composite connections with screws and toothed plates. *Materials* **2021**, *15*, 68. [CrossRef]
37. Raftery, G.M.; Whelan, C. Low-grade glued laminated timber beams reinforced using improved arrangements of bonded-in GFRP rods. *Constr. Build. Mater.* **2014**, *52*, 209–220. [CrossRef]
38. Gómez, S.; Svecova, D. Behavior of split timber stringers reinforced with external GFRP sheets. *J. Compos. Constr.* **2008**, *12*, 202–211. [CrossRef]
39. Hay, S.; Thiessen, K.; Svecova, D.; Bakht, B. Effectiveness of GFRP sheets for shear strengthening of timber. *J. Compos. Constr.* **2006**, *10*, 483–491. [CrossRef]
40. Wu, C.; Zhang, Z.; Tam, L.-h.; Feng, P.; He, L. Group effect of GFRP-timber bolted connections in tension. *Compos. Struct.* **2021**, *262*, 113637. [CrossRef]
41. Cabral, J.P.; Kafle, B.; Subhani, M.; Reiner, J.; Ashraf, M. Densification of timber: A review on the process, material properties, and application. *J. Wood Sci.* **2022**, *68*, 20. [CrossRef]
42. Feng, T.Y.; Chiang, L.K. Effects of densification on low-density plantation species for cross-laminated timber. In Proceedings of the AIP Conference Proceedings, Kuala Lumpur, Malaysia, 26–28 August 2019.
43. Salca, E.-A.; Bekhta, P.; Seblii, Y. The effect of veneer densification temperature and wood species on the plywood properties made from alternate layers of densified and non-densified veneers. *Forests* **2020**, *11*, 700. [CrossRef]
44. Kennedy, S.; Salenikovich, A.; Munoz, W.; Mohammad, M. Design equations for dowel embedment strength and withdrawal resistance for threaded fasteners in CLT. In Proceedings of the World Conference on Timber Engineering, Quebec, QC, Canada, 10–14 August 2014; pp. 10–14.
45. National Design Specification (NDS). *National Design Specification (NDS) for Wood Construction–with Commentary*, 2015th ed.; American Wood Council: Leesburg, VA, USA, 2014.
46. Uibel, T.; Blaß, H.J. Load carrying capacity of joints with dowel type fasteners in solid wood panels. In Proceedings of the CIB-W18 Meeting, Bled, Slovenia, 8 August 2007.
47. *CSA. O86-14*; Engineering Design in Wood. Canadian Standards Association: Mississauga, ON, Canada, 2014.
48. Long, W.; Ou, J.; Sun, X.; Huang, X.; He, M.; Li, Z. Experimental study on the embedment strength of smooth dowels inserted in cross-laminated timber narrow side. *J. Wood Sci.* **2022**, *68*, 1–18. [CrossRef]
49. Tuhkanen, E.; Mölder, J.; Schickhofer, G. Influence of number of layers on embedment strength of dowel-type connections for glulam and cross-laminated timber. *Eng. Struct.* **2018**, *176*, 361–368. [CrossRef]
50. Sawata, K.; Yasumura, M. Determination of embedding strength of wood for dowel-type fasteners. *J. Wood Sci.* **2002**, *48*, 138–146. [CrossRef]
51. Lewis, C.D. *Industrial And Business Forecasting Methods: A Practical Guide to Exponential Smoothing and Curve Fitting*; Butterworth Scientific: Oxford, UK, 1982.
52. Nasir, V.; Ayanleye, S.; Kazemirad, S.; Sassani, F.; Adamopoulos, S. Acoustic emission monitoring of wood materials and timber structures: A critical review. *Constr. Build. Mater.* **2022**, *350*, 128877. [CrossRef]

Disclaimer/Publisher's Note: The statements, opinions and data contained in all publications are solely those of the individual author(s) and contributor(s) and not of MDPI and/or the editor(s). MDPI and/or the editor(s) disclaim responsibility for any injury to people or property resulting from any ideas, methods, instructions or products referred to in the content.

Article

Enhancing Surface Characteristics and Combustion Behavior of Black Poplar Wood through Varied Impregnation Techniques

Abdullah Beram

Department of Industrial Design, Faculty of Architecture and Design, Pamukkale University, 20160 Denizli, Türkiye; abdullahberam@pau.edu.tr

Abstract: The objective of this work was to improve the thermal stability, flame resistance, and surface properties of black poplar (*Populus nigra* L.) wood via different impregnation methods. The impregnation methods were employed through two distinct modalities: vacuum impregnation and immersion impregnation. Here, poplar wood was impregnated with calcium oxide solutions (1%, 3% and 5%). Fourier-transform infrared spectroscopic analysis revealed a shift in the typical peaks of cellulose, hemicellulose, and lignin depending on the impregnation method and solution ratio. Thermogravimetric analysis and the limiting oxygen index indicated that the samples impregnated with lime solutions exhibited higher thermal stability than the unimpregnated wood. Both impregnation methods caused a decrease in water absorption and thickness swelling of the sample groups. Using a scanning electron microscope, the effect of the impregnation process on the structure of the wood was examined. In terms of surface properties, it was determined that the surface roughness value increased. On the contrary, it was observed that the contact angle value also increased. A significant difference emerged between the applied methods. In conclusion, the applied lime minerals are suitable substances to increase the flame resistance and thermal stability of black poplar wood.

Keywords: black poplar; impregnation; surface properties; thermal stability; wood protection

Citation: Beram, A. Enhancing Surface Characteristics and Combustion Behavior of Black Poplar Wood through Varied Impregnation Techniques. *Appl. Sci.* **2023**, *13*, 11482. https://doi.org/10.3390/app132011482

Academic Editors: Alena Očkajová, Martin Kučerka and Richard Kminiak

Received: 20 September 2023
Revised: 17 October 2023
Accepted: 17 October 2023
Published: 19 October 2023

Copyright: © 2023 by the author. Licensee MDPI, Basel, Switzerland. This article is an open access article distributed under the terms and conditions of the Creative Commons Attribution (CC BY) license (https://creativecommons.org/licenses/by/4.0/).

1. Introduction

Wood, a fundamental natural resource, has been employed in various applications since time immemorial. Its versatility, renewable nature, and wide availability have rendered it an indispensable material in industries such as construction, furniture, and manufacturing [1–3]. However, one of the perennial challenges associated with wood products is their susceptibility to combustion. The threat of fire not only poses safety concerns but also has substantial economic and ecological ramifications. Hence, there has been an enduring quest to enhance the fire-resistant properties of wood [4–10].

Among the numerous wood species available, poplar wood (*Populus* spp.) stands out for its fast growth rate and ease of cultivation. Poplar is widely distributed across temperate regions, and its utilization has surged in recent years, particularly in applications where rapid growth is essential [11]. Black poplar (*Populus nigra* L.) is one of these species. Black poplar is distributed in North Africa, Central and Western Asia, and Europe, especially in wetlands along riverbanks [12]. The distribution of poplar species in the world is more than 100 million ha. Türkiye ranks fourth in the world in terms of poplar plantation area [13]. More than 3 million m³ of wood are obtained annually from this species alone in Türkiye [14]. Although black poplar wood is widely used in furniture production, it can also find a place as a raw material in the packaging industry (boxes, crates, pallets, etc.) and in the production of models, plywood, matches, composite panels, and prostheses [15,16]. Nevertheless, like many other wood varieties, poplar wood is inherently vulnerable to fire, necessitating innovative approaches to improve its fire-resistant characteristics.

In this context, the impregnation of wood with fire-retardant chemicals has emerged as a promising avenue for enhancing its fire resistance. Impregnation involves the penetration of wood with fire-retardant substances, which can alter the wood's surface properties and combustion behavior [17–19]. The choice of impregnation technique and the type of fire retardant used are important factors in determining the effectiveness of this process. Therefore, it becomes imperative to explore the influence of diverse impregnation techniques on the surface characteristics and combustion behavior of poplar wood [20–25].

The practice of impregnating wood with various substances to enhance its properties is a method that has been used for many years. It has been employed for centuries, albeit with rudimentary techniques. Modern wood impregnation techniques have evolved significantly [26–30]. One of the earliest methods involved simply soaking wood in a solution containing fire-retardant chemicals. While this method is straightforward, it often results in uneven impregnation and inadequate penetration of fire retardants into the wood's cellular structure. To address these limitations, vacuum impregnation and pressure impregnation techniques were developed. Vacuum impregnation, in particular, involves subjecting wood to reduced pressure before immersing it in a fire-retardant solution [24,31–34].

The choice of fire-retardant chemicals is a critical determinant of the efficacy of wood impregnation. Fire retardants can be categorized into several classes, including inorganic compounds, organic compounds, and intumescent agents [35–38]. Inorganic fire retardants, such as ammonium phosphate and aluminum hydroxide, work by releasing water vapor when exposed to heat, thereby reducing the wood's temperature and retarding combustion. One of these, calcium oxide (CaO), is a white, corrosive, and alkaline solid [39]. Calcium oxide is used in the construction industry and in the production of paper, among many other applications, such as the manufacture of various types of glass [39,40]. These compounds are known for their non-toxic nature and widespread use in wood impregnation.

In this study, black poplar wood was subjected to different impregnation methods with calcium hydroxide in order to improve its physical properties and resistance to burning. Solutions prepared at different concentrations (1%, 3% and 5%) were used in the vacuum method and the immersion method. The chemical and thermal changes caused by the impregnation process in the samples were evaluated by comparison with the control samples. The effects of different impregnation methods at different durations and concentrations on the physical and fire properties of the samples were investigated.

2. Materials and Methods

2.1. Materials

2.1.1. Wood Material

The material of black poplar used in the study was obtained from Denizli Kırgız Timber Company, Türkiye. According to TS 2470 standards [41], the samples were made from sapwood and first-class timber materials that are smooth fiber, knotless, crack-free, without color or density differences, and with yearly rings perpendicular to the surfaces. A total of 30 samples were taken for each experimental group.

2.1.2. Impregnation Material (Calcium Hydroxide ($Ca(OH)_2$))

Calcium oxide (CaO), often known as quicklime, is a substance that is frequently utilized as a commercial product. At room temperature, it is an alkaline solid that is white and caustic. In addition to CaO, quicklime contains magnesium oxide (MgO) and silicon dioxide (SiO_2), with minor traces of aluminum oxide (Al_2O_3) and iron oxide (Fe_2O_3). This phenomenon is attributed to the inherent presence of these contaminants in the raw material, 'limestone' [42,43] (Table 1). CaO powder was purchased from Ayteks Chemical Industry Ltd. Denizli/Türkiye. Prior to its application in the impregnation process, calcium oxide necessitates slaking (adding water to the lime). Afterwards, the powdered lime was weighed to 1%, 3%, and 5% (w/v); the solutions were prepared through the slaking of lime to Equation (1) [43].

Table 1. Identifiers and properties of calcium oxide [43].

CAS number	1305-78-8
PubChem CID	14778
UN number	1910
Molecule formula	CaO
Molecular mass	56.0774 g/mol
Appearance	White to yellow/brown powder
Odor	odorless
Density	3.34 gr/cm^3
Melting point	2613 °C
Boiling point	3850 °C (100 hPa)
Solubility	(in water) reacts to form calcium hydroxide
Acidity (pKa)	12.8

Calcium hydroxide exhibits a relatively low water solubility. Investigating its solubility, it was determined [44] that it amounts to 0.0222 molal at 25 °C, corresponding to a low solubility of 1.6 g in 1 kg of water.

$$CaO + H_2O \rightarrow Ca(OH)_2 \quad (1)$$

2.2. Methods

2.2.1. Impregnation Methods (Immersion Method and Vacuum Method)

Calcium hydroxide ($Ca(OH)_2$) solutions at concentrations of 1%, 3%, and 5% (w/v) were prepared. According to the ASTM D 1413 standard, samples were impregnated with these solutions using the medium-term (120 min) immersion method.

The vacuum method of impregnation of the test samples was carried out under the conditions specified in ASTM D 1413. In this impregnation process, the samples were subjected to pre-vacuum treatment at 760 mmHg^{-1} with a compressor in a closed container for 30 min. Then, the samples were removed from the closed container and impregnated in solution at atmospheric pressure for 30 min [45].

The samples in wet weights were held in the air-conditioning cabinet at a temperature of 20 ± 2 °C and a relative humidity of 65 ± 5% until they reached equilibrium humidity after impregnations. After being impregnated, the samples were maintained in an oven set at 103 ± 2 °C until they were entirely dry. After these steps, the characterization was conducted. The experimental design of the samples used in the study is given in Table 2.

Table 2. Experimental design of immersion and vacuum methods.

Method	Sample Type	Concentration of Solution (%)	Impregnation Time (min)
Control	A	-	-
Immersion	B	1	120
	C	3	120
	D	5	120
Vacuum	E	1	30 + 30
	F	3	30 + 30
	G	5	30 + 30

2.2.2. Preparation of Samples

Density (D), thickness swelling (TS) and water absorption (WA) experiments were carried out with samples of 20 × 20 × 30 mm^3 (L × T × R) volume. The density, dimensional change, and water uptake determinations of the samples were conducted in compliance with TS 2472, TS 4084, and TS EN 317 standards, respectively [46–48]. For each treated wood sample, the weight percent gain (WPG) was calculated according to Equation (2). The oven-dry weight of each specimen was recorded before and after impregnation. The assessment of surface and fire properties adhered to relevant criteria. The residues left on the wood surface from the impregnation process were removed before the tests to ensure they did not affect the results.

$$\text{WPG} = \frac{(Wa - Wb)}{Wb} \times 100\% \tag{2}$$

where W_b is the oven-dry weight of specimens before treatment (g), and W_a is the oven-dry weight of specimens after treatment (g).

2.2.3. Testings of Surface and Burning Properties

- Surface roughness test

Surface roughness measurements were conducted in accordance with the DIN 4768 (1990) standard [49], employing a stylus-type profilometer (Mitutoyo SJ-301, Mitutoyo Corp., Kawasaki, Japan). The measurement was taken on the wood surfaces in parallel to the grain direction ($\parallel$). The roughness values were captured with a sensitivity of 0.5 μm, ensuring precision in the measurements. The key instrument parameters included a measuring speed of 10 mm/min, a pin diameter of 4 μm, and a pin top angle set at 90 degrees. The points selected for roughness measurement were deliberately marked in a random manner across the sample surface. All measurements were conducted parallel to the direction of the wood fibers. Three standard roughness parameters were determined: average roughness (Ra); mean peak-to-valley height (Rz); and maximum roughness (Rmax). Ra values were specifically employed for statistical evaluations. Additionally, measurements were repeated whenever the scanning needle's tip entered the cell spaces within the wood sample.

- Water contact angle test

According to GB/T 30693 (2014), the water contact angle of the surface of wood was calculated. KRUSS DSA30 water contact angle meter (KRÜSS, Hamburg, Germany) was used. The size of the water drop was 4 μL. The data were obtained 15 s after the water droplet made contact with the wood surface. The contact angles were obtained at five separate sites on the same sample surface using five replicates for each group, and the mean value was calculated.

- FTIR analysis

The samples were finely ground, falling within the 40–100 mesh size range, in preparation for their utilization in Fourier-transform infrared (FTIR) spectroscopy and thermogravimetric analyses (TGA). Following the grinding process, pellets were created by subjecting 10 mg of wood flour and KBr to a 1:100 (w/w) ratio for each sample group. These pellets were formed by applying a pressure of 602 N/mm^2. For the FTIR analysis, a Perkin Elmer BX FTIR spectrometer instrument (PerkinElmer U.S. LLC, Shelton, CT, USA) was employed, operating at ambient temperature. The instrument covered a wavenumber range of 4000 to 400 cm^{-1}, with a spectral resolution of 4 cm^{-1}.

- DTG/TGA analysis

To assess the thermal stability of the samples, thermogravimetric analysis (TGA) was conducted using an Exstar SII TG DTA 7200 (Exstar, SII NanoTechnology In., Tokyo, Japan) instrument. The analysis was carried out under a nitrogen gas atmosphere, with the samples experiencing a gradual temperature increase at a rate of 10 °C per minute, spanning from 25 to 600 °C. Each sample was weighed at approximately 5 mg.

- LOI analysis

A flammability test to determine the limiting oxygen index (LOI) of wood samples was carried out using a flammability tester (S.C. Dey Co., Kolkata, India). This test was conducted in accordance with the ASTM D-2863 method [50]. The LOI value expresses the minimum amount of oxygen required to start combustion. In the LOI apparatus, the wood sample was positioned vertically and subjected to ignition for a minimum duration of 30 s. Throughout the ignition process, the ratio between nitrogen and oxygen in the environment was carefully monitored and recorded.

- SEM analysis

A scanning electron microscope (FESEM, Zeiss Gemini Sigma 300, Jena, Germany) equipped with an energy dispersive spectrometer (EDS) system (Bruker XFlash 6I100) was employed to reveal the effect of impregnation on the particles.

- Statistical analysis

Statistical analysis was performed on the study's findings using SPSS® 20.0 for Windows® software. The data were subjected to an analysis of variance (ANOVA) test. A Duncan test was used to identify the various groups in cases where the ANOVA test revealed statistical differences via SPSS® 20.0 for Windows® (IBM Corp., Armonk, NY, USA).

3. Results and Discussion

- The physical properties

The physical characteristics of the samples of black poplar with different impregnation methods are presented in Table 3. Using the ANOVA test, it was found that there was a statistically significant difference between the control and impregnated sample sets in terms of the physical characteristics of the experimental specimens. After applying the Duncan test, four homogeneous clusters were delineated within each of the datasets corresponding to D_0, TS-2, and TS-24 h and five homogeneous clusters were delineated within each of the datasets corresponding to WA-2, WA-24, and WPG.

Table 3. The physical properties of impregnated wood samples.

Sample Type	D_0 (g/cm^3)	WA-2 h	WA-24 h	TS-2 h	TS-24 h	WPG (%)
A	0.36 (0.09) [1] a [2]	38.21 (3.57) a	72.89 (5.88) a	14.86 (2.43) a	17.26 (3.45) a	-
B	0.41 (0.13) b	33.83 (3.26) b	67.86 (6.11) b	13.23 (1.31) b	15.08 (3.04) b	0.62 (0.09) a
C	0.44 (0.09) c	29.42 (2.61) c	63.22 (5.67) c	11.38 (1.14) c	13.77 (2.14) c	0.73 (0.21) b
D	0.45 (0.11) c	26.55 (1.93) d	59.49 (4.93) d	10.75 (0.81) cd	12.13 (1.86) cd	0.91 (0.18) b
E	0.42 (0.12) b	31.26 (2.79) c	63.92 (4.55) c	11.49 (1.05) c	14.86 (2.11) b	1.12 (0.30) c
F	0.46 (0.14) c	24.12 (2.44) d	58.32 (5.22) d	9.08 (1.27) d	12.22 (1.37) cd	1.79 (0.35) d
G	0.49 (0.09) d	22.07 (1.66) e	54.11 (4.83) e	8.83 (0.95) d	11.45 (1.96) d	2.28 (0.51) e

[1]: Standard deviation, [2]: The same letters in a column of D0, WA-2, WA-24, TS-2, TS-24, and WPG are not significantly different ($p \leq 0.01$). The post hoc tests (Duncan) for the D0, WA-2, WA-24, TS-2, TS-24, and WPG were analyzed separately because the interaction factor was significantly different.

It was found that the density (D_0) and weight percent gain (WPG) rose when the lime ratio increased in the two impregnation methods. When 1%, 3%, and 5% lime were added, respectively, the D_0 values were found to be between 0.41 and 0.49 g/cm^3. Depending on this, the WPG increased between 0.62% and 2.28%. Applied impregnation methods with lime progressively decreased water absorption and thickness swelling in the samples. It was observed that WA-2, WA-24, TS-2, and TS-24 values decreased when 1%, 3%, and 5% lime were added, respectively.

The WA-2 decreased between 11.5% and 30.5%, the WA-24 values decreased between 6.9% and 18.4%, the TS-2 values decreased between 10.9% and 27.6%, and the TS-24 values decreased between 12.6% and 29.7% in the immersion method. In the vacuum method, which is the other method applied, the WA-2 decreased between 18.2% and 42.2%, WA-24 values decreased between 12.3% and 25.7%, TS-2 values decreased between 22.6% and 40.5%, and TS-24 values decreased between 13.9% and 33.6% (Table 3). These data are consistent with earlier research, which found that adding lime to wood increased its physical qualities and made it more stable dimensionally [5,8,9,33]. In addition, there appear to be obvious differences in physical properties between the applied methods. It is seen that the vacuum method provides more stability to the wood material compared to the immersion method [51–53].

- The surface roughness, contact angle, and LOI properties

Table 4 displays the surface roughness, contact angle, and LOI characteristics of samples of black poplar treated with various impregnation methods. According to the ANOVA test, a statistically significant difference was detected in the physical properties of both control and impregnated sample groups. Following the implementation of the Duncan test, we identified four consistent and similar groups within each dataset associated with surface roughness, contact angle, and LOI.

Table 4. Surface roughness, contact angle, and LOI properties of impregnated wood samples.

Sample Type	Surface Roughness (Ra) ($\|$)	Changes (%)	Contact Angle (°)	Changes (%)	LOI (%)	Changes (%)
A	2.77 (0.31)[1] a [2]	-	41 (4.66) a	-	23.16 (2.55) b	-
B	3.36 (0.99) b	21.3	54 (4.34) b	31.7	26.75 (3.07) b	15.5
C	3.90 (0.83) c	40.8	59 (5.27) c	43.9	28.44 (2.43) b	22.8
D	4.12 (0.46) d	48.7	62 (4.02) d	51.2	30.08 (1.94) b	29.8
E	3.58 (0.60) c	29.2	61 (3.95) c	48.8	28.27 (1.64) b	22.0
F	4.35 (0.97) d	57.0	66 (6.26) e	60.9	30.62 (2.44) b	32.2
G	5.22 (0.44) e	88.4	68 (6.31) e	65.8	31.23 (2.74) b	34.8

[1]: Standard deviation, [2]: The same letters in a column of surface roughness, contact angle, and LOI are not significantly different ($p \leq 0.01$). The post hoc tests (Duncan) for the surface roughness, contact angle, and LOI were analyzed separately because the interaction factor was significantly different.

The average roughness parameter (Ra) increased with an increase in the solution ratio. The values were found to be between 2.77 and 5.22. Compared to the control group, the B group exhibited the smallest alteration, registering a 21.3% change, while the G group displayed the most substantial variation with an 88.4% shift. Ra increases the surface roughness of the impregnation process. It is explained that this situation is related to the increase in the amount of substance on the surface as the amount of retention increases [54–56].

The contact angle values of the groups included in the study are shown in Table 4. It has been determined that as the lime concentration ratio increases, the contact angle increases. The values were found to be between 41° and 68°. It has been determined that the highest hydrophobic sample group with a contact angle of 68° is obtained with group G, which increases hydrophobicity by 65.8% compared to group A. Similar to the findings for water absorption, increasing lime particles increased the contact angle, which was significantly higher in woods treated and impregnated [57–60].

The LOI values of the sample groups are summarized in Table 4. The values were found to be between 23.16% and 31.23%. It has been determined that the highest fireproof sample group with a LOI of 31.23% is obtained with group G, which increases fire resistance by 34.8% compared to group A. In wood impregnated with lime minerals, the LOI value increased as the lime ratio increased. LOI values were found to be between 26.75% and 30.08% in the immersion method and between 28.27% and 31.23% in the vacuum method. The vacuum method is posited to yield a superior insulating effect against heat transfer compared to the immersion method. The retardation of flame propagation appears to stem from the lime's capacity to facilitate the generation of char. This ensuing coal coating forms an insulative barrier, impeding the passage of combustible gases that sustain the flame and displaying resistance to heat transfer [61–63].

- FTIR analysis

FTIR analysis was employed to discern the functional groups and chemical interactions among the materials. The FTIR spectrum exhibited observable shifts in the characteristic peaks of cellulose, hemicellulose, and lignin, contingent upon the impregnation method and the ratio of lime additive. FTIR spectra encompassing the impregnated black poplar samples, as well as the control samples, were recorded over the wavelength range of 4000 to 400 cm^{-1}. The control group and the groups impregnated with both impregnation methods show absorbance peaks for wood fibers at 876 cm^{-1} (Si–O–C), 1060 cm^{-1} (C–O–C),

2910 cm^{-1} (C–H), and 3450 cm^{-1} (O–H). In addition, the impregnated groups showed a new increase at 1450 cm^{-1} for C=O stretching vibration. The lack of a drop in the strength of the band at 1060 cm^{-1} indicates that the cellulose's C–O–C bonds have not been harmed by the procedure. It may be argued that the impregnated lime particles are to blame for this phenomenon (Table 5 and Figure 1). The absorption bands over 1450 cm^{-1} can be assigned to the stretching vibrations of (CO$_3$)- anions present in the carbonate phase in the sample. The behavior of Ca(OH)$_2$ adsorbed on the surface was monitored, showing that Ca(OH)$_2$ continuously transformed into the carbonate phase and crystallization proceeded first through the formation of aragonite-like and then calcium-like carbonates [64].

Table 5. Match of wood functional groups to IR bands of spectra [65,66].

Spectrum Band Position, cm^{-1}	Active Wood Mass Group	Type of Vibration
3450–3400	O-H of alcohols, phenols and acids	O-H stretching
2970–2850	CH$_2$, CH- and CH$_3$	C-H stretching
1462–1425	CH$_2$ cellulose, lignin	C-H deformations
1060–1025	C-O-C	Deformation
876	Anti-symmetric out-of-phase stretching in pyranose ring	Stretching in pyranose ring

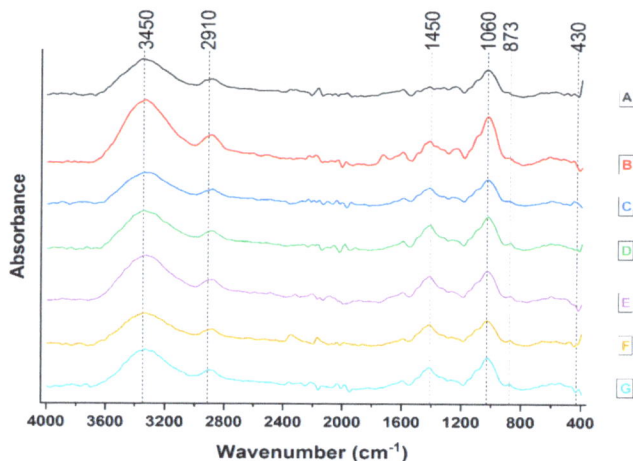

Figure 1. FTIR Spectra of the impregnated and unimpregnated black poplar samples.

The presence of a band at 3450 cm^{-1} signifies a reduction in the quantity of OH groups, leading to a further decrease compared to the control group. An examination of the FTIR spectroscopy peaks reveals notable alterations in cellulose, hemicellulose, and lignin due to the processing [23,66,67]. In contrast to the control group, the peak observed at 2910 cm^{-1} is notably diminished. This decrease can be attributed to the asymmetric stretching of C–H methyl and methylene groups [68–70]. Conversely, a noticeable increase is evident in the peak at 1450 cm^{-1} compared to the control group. This peak is characteristic of lignin components and signifies symmetrical tension vibrations in C=O and –COO groups within aromatic rings [71–73]. Moreover, these changes elucidate the influence of functional groups in the added lime minerals on the wood. Another significant observation is the increase in the 873 cm^{-1} band relative to the control group, which can be attributed to the Si–O–Al stretching mode associated with CaO [74,75]. Additionally, distinct peak bands are discernible at 430 cm^{-1} [23,76]. Those findings suggest that lime minerals were successfully grafted into the poplar wood fibers.

- DTG/TGA analysis

The TGA and DTG thermograms of the control and impregnated black poplar samples are plotted in Figures 2 and 3.

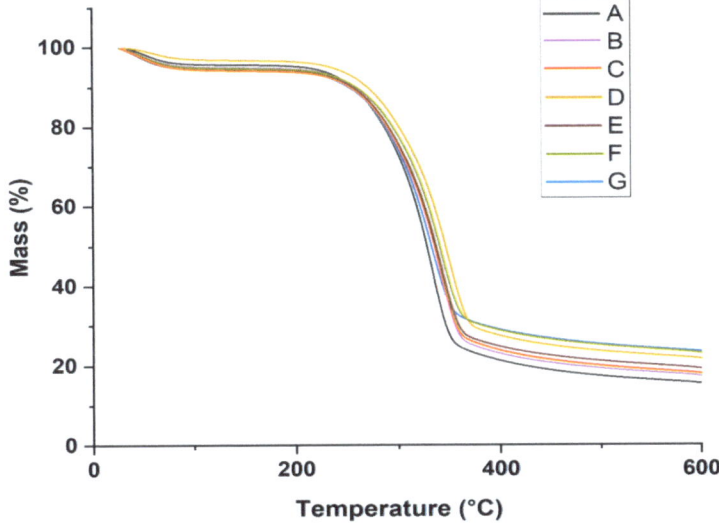

Figure 2. TGA thermograms of the control and impregnated black poplar samples.

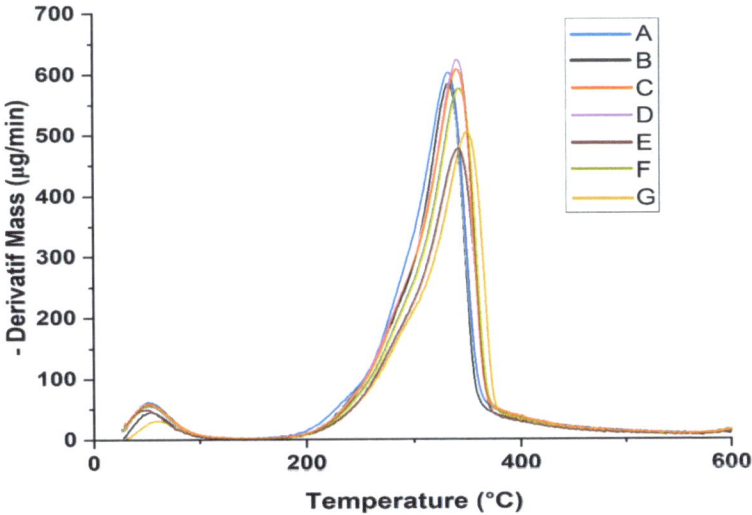

Figure 3. DTG thermograms of the control and impregnated black poplar samples.

The provided data in Table 6 summarize the initial decomposition temperature (T_0), maximum degradation temperature (T_{max}), final temperature (T_f), and residual weight (RW, %) for the wood samples, both impregnated and control, with calcium hydroxide. The onset of degradation occurred at 140 °C for both the impregnated and non-impregnated black poplar samples, signifying the removal of water and certain extractive components from the specimens up to this temperature [77,78]. After 140 °C, the decomposition process continued until 476 °C in the control sample, between 494 and 531 °C in the samples impregnated with the immersion method, and between 532 and 584 °C in the samples impregnated with the vacuum method. The highest final temperature was determined in the G sample at 584 °C. The maximum degradation temperature is the lowest value in the control sample at 329 °C and the highest value in the G sample at 347 °C. From 140 °C to 476 °C, and 584 °C, hemicellulose, the remaining extractives, lignin, and cellulose were decomposed [22,79]. The residue weight varied depending on the method of impregnation. The rate of RW at 600 °C in the samples was 16.2% in the control sample (A), between 18.3 and 22.3% in the samples impregnated with the immersion method, and between 19.8 and 24.9% in the samples impregnated with the vacuum method. The TGA study results showed that as the concentration of calcium hydroxide increased, the heat resistance of the fibers gradually increased. Additionally, the amount of residue detected in the vacuum method is slightly higher than in the immersion method. These values are relatively low when compared with the literature results [80–84].

Table 6. Thermal degradation temperatures and residue weight of black polar samples.

Sample Type	T_0 (°C)	T_{max} (°C)	T_f (°C)	RW at 600 °C (%)
A	140	329	476	16.2
B	140	332	494	18.3
C	140	337	503	18.5
D	140	338	531	22.3
E	140	339	532	19.8
F	140	342	576	24.4
G	140	347	584	24.9

T_0: Initial decomposition temperature, T_{max}: maximum degradation temperature, T_f: Final temperature.

- SEM analysis

Observation under SEM at high magnifications showed the samples of impregnated and unimpregnated black poplar (Figures 4 and 5).

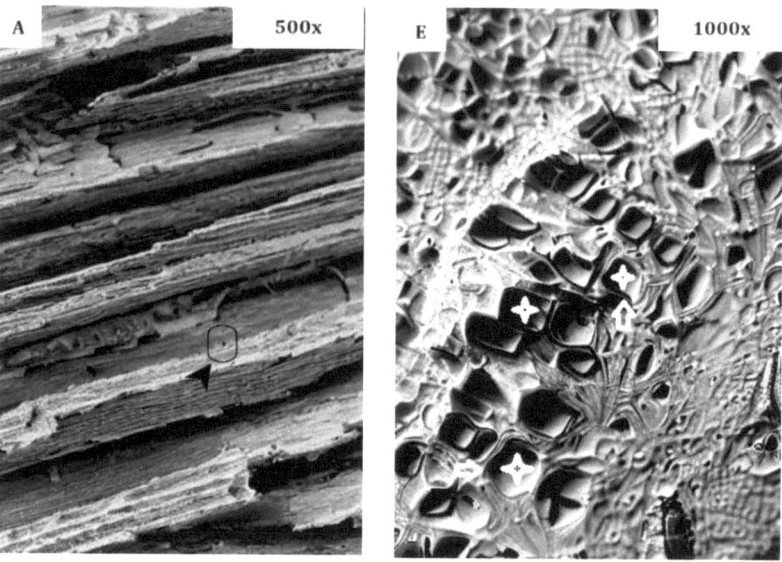

Figure 4. SEM topographs of black poplar: (A) control (500×); (E) impregnated sample (1000×).

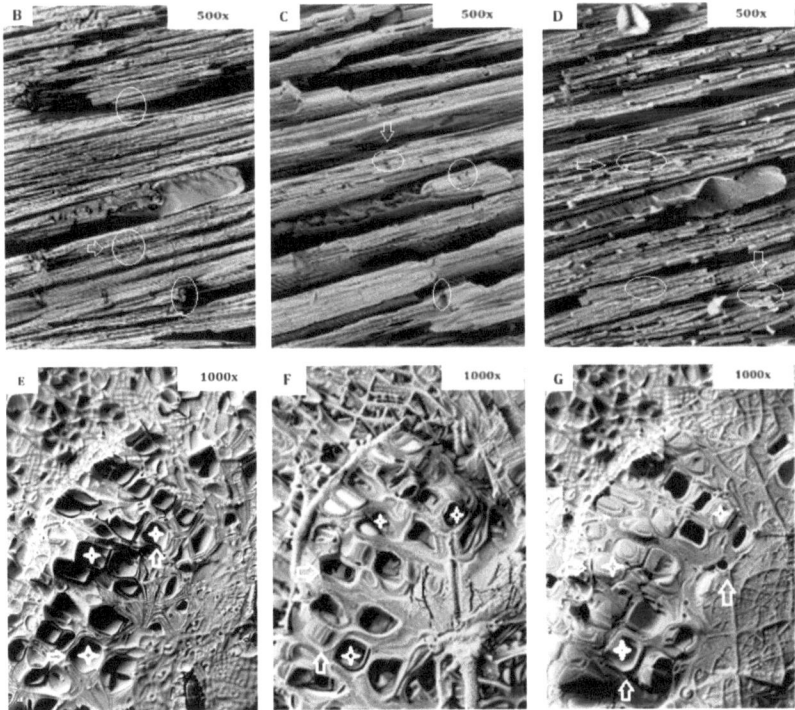

Figure 5. SEM micrographs of impregnated samples of black poplar: (B–D) (500×), (E–G) (1000×).

SEM analysis of impregnated wood material revealed the presence of impregnation substances concentrated along the wood lumen cell and transition edges. Additionally, nanoparticles were observed to form clusters within certain regions of the trachea [85]. It can be seen that the amount of impregnation filling the lumen cell is related to the change in concentration.

4. Conclusions

This study was undertaken to enhance the surface characteristics and fire-resistant properties of black poplar wood, a rapidly growing tree species. In pursuit of this objective, two distinct methods were employed for the impregnation of calcium hydroxide. In comparison to the immersion method, the vacuum impregnation method produced better results, showing lower water absorption and thickness swelling values. This resulted in an increase in hydrophobic characteristics of the wood. Notably, an increase in the weight percent gain (WPG) ratio corresponded with a successful impregnation process and a concurrent elevation in limiting oxygen index (LOI) values, suggesting improved fire resistance. The Fourier-transform infrared (FTIR) analysis findings aligned with the thermogravimetric analysis-differential thermal gravimetry (TGA-DTG) results, demonstrating an augmentation in residue content as the concentration rate of impregnation increased. These analyses affirm enhanced fireproof properties. Examination of scanning electron microscopy (SEM) images revealed some deposits in lumen cell occupancy, indicative of a successful impregnation process. Additionally, it was observed that while porosity decreased, surface roughness increased due to the disintegration of structural elements. This effect, however, led to an increase in contact angle values and the filling of surface gaps on the poplar wood. In conclusion, calcium hydroxide emerges as an auspicious candidate for augmenting the fire-resistant attributes of wood materials. The applicability of this approach should be considered in accordance with specific use cases and the structural limitations inherent to poplar wood, thereby facilitating the production of more efficacious end products.

Funding: This research received no external funding.

Institutional Review Board Statement: Not applicable.

Informed Consent Statement: Not applicable.

Data Availability Statement: Not applicable.

Conflicts of Interest: The author declares no conflict of interest.

References

1. Asif, M. Sustainability of timber, wood and bamboo in construction. In *Sustainability of Construction Materials*; Khatib, J.M., Ed.; Woodhead Publishing: Oxford, UK, 2009; pp. 41–54. [CrossRef]
2. Rosillo-Calle, F.; Woods, J. *The Biomass Assessment Handbook*; Routledge: London, UK, 2012.
3. Aristri, M.A.; Lubis, M.A.R.; Yadav, S.M.; Antov, P.; Papadopoulos, A.N.; Pizzi, A.; Fatriasari, W.; Ismayati, M.; Iswanto, A.H. Recent developments in lignin-and tannin-based non-isocyanate polyurethane resins for wood adhesives—A review. *Appl. Sci.* **2021**, *11*, 4242. [CrossRef]
4. Slimak, K.M.; Slimak, R.A. Enhancing the Strength, Moisture Resistance, and Fire-Resistance of Wood, Timber, Lumber, Similar Plant-Derived Construction and Building Materials, and Other Cellulosic Materials. U.S. Patent No. 6,146,766, 21 March 2000.
5. Östman, B.; Voss, A.; Hughes, A.; Jostein Hovde, P.; Grexa, O. Durability of fire-retardant treated wood products at humid and exterior conditions review of literature. *Fire Mater.* **2001**, *25*, 95–104. [CrossRef]
6. Pereyra, A.M.; Giudice, C.A. Flame-retardant impregnants for woods based on alkaline silicates. *Fire Saf. J.* **2009**, *44*, 497–503. [CrossRef]
7. Wen, M.Y.; Kang, C.W.; Park, H.J. Impregnation and mechanical properties of three softwoods treated with a new fire-retardant chemical. *J. Wood Sci.* **2014**, *60*, 367–375. [CrossRef]
8. Fu, Q.; Medina, L.; Li, Y.; Carosio, F.; Hajian, A.; Berglund, L.A. Nanostructured wood hybrids for fire-retardancy prepared by clay impregnation into the cell wall. *ACS Appl. Mater. Interfaces* **2017**, *9*, 36154–36163. [CrossRef] [PubMed]
9. Madyaratri, E.W.; Ridho, M.R.; Aristri, M.A.; Lubis, M.A.R.; Iswanto, A.H.; Nawawi, D.S.; Antov, P.; Kristak, L.; Majlingová, A.; Fatriasari, W. Recent advances in the development of fire-resistant biocomposites—A review. *Polymers* **2022**, *14*, 362. [CrossRef]
10. Yu, Z.L.; Ma, Z.Y.; Yao, H.X.; Qin, B.; Gao, Y.C.; Xia, Z.J.; Huang, Z.H.; Yin, Y.C.; Tu, H.; Ye, H.; et al. Economical architected foamy aerogel coating for energy conservation and flame resistance. *ACS Mater. Lett.* **2022**, *4*, 1453–1461. [CrossRef]
11. Ercan, M. *Poplar Research Institute from its Establishment to the Present 1962–2014*; (T.R. Ministry of Forestry and Water Affairs, General Directorate of Forestry, Poplar and Fast-Growing Forest Trees Research Institute. Directorate Publication No: 270); Various Publications Series No: 25: Izmit, Türkiye, 2014. Available online: https://kutuphane.tarimorman.gov.tr/vufind/Record/10258 (accessed on 5 August 2023).
12. Marchi, M.; Bergante, S.; Ray, D.; Barbetti, R.; Facciotto, G.; Chiarabaglio Pier, M.; Nervo, G. Universal reaction norms for the sustainable cultivation of hybrid poplar clones under climate change in Italy. *Iforest Biogeosciences For.* **2022**, *15*, 47. [CrossRef]
13. Atmaca, C. Performance of Various Poplar Clones in the Early Years. Master's Thesis, Düzce University, Institute of Science and Technology, Düzce, Türkiye, 2018.
14. Birler, A.S. *Poplar Cultivation in Türkiye: Nursery-Afforestation-Protection-Revenue-Economy-Wood Characteristics*; Poplar and Fast-Growing Forest Trees Research Directorate of the Ministry of Environment and Forestry: Ankara, Türkiye, 2010.
15. Bozkurt, Y.; Erdin, N. *Wood Anatomy*; İstanbul University, Publishing of Faculty of Forestry: Istanbul, Türkiye, 2000.
16. Gaudet, M.; Jorge, V.; Paolucci, I.; Beritognolo, I.; Scarascia Mugnozza, G.; Sabatti, M. Genetic linkage maps of *Populus nigra* L. including AFLPs, SSRs, SNPs, and sex trait. *Tree Genet. Genomes* **2008**, *4*, 25–36. [CrossRef]
17. Uysal, B.; Yapıcı, F.; Kol, H.Ş.; Özcan, C.; Esen, R.; Korkmaz, M. Determination of thermal conductivity finished on impregnated wood material. In Proceedings of 6th International Advanced Technologies Symposium (IATS'11), Elazığ, Türkiye, 6–18 May 2011.
18. Kesik, H.İ.; Keskin, H.; Temel, F.; Öztürk, Y. Bonding Strength and Surface Roughness Properties of Wood Materials Impregnated with VacsolAqua. *Kastamonu Univ. J. For. Fac.* **2016**, *16*, 181–189. [CrossRef]
19. Demir, A.; Aydin, İ. Effects of Treatment with Fire Retardant Chemicals on Technologic Properties of Wood and Wooden Materials. *Duzce Univ. Fac. For. J. For.* **2016**, *12*, 96–104.
20. Göker, Y.; Ayrılmış, N. Performance characteristicsand thermal degradation of wood and wood-based products in fire. *J. Fac. For. Istanb. Univ.* **2002**, *52*, 1–22.
21. He, X.; Li, X.J.; Zhong, Z.; Mou, Q.; Yan, Y.; Chen, H.; Liu, L. Effectiveness of impregnation of ammonium polyphosphate fire retardant in poplar wood using microwave heating. *Fire Mater.* **2016**, *40*, 818–825. [CrossRef]
22. Kong, L.; Guan, H.; Wang, X. In situ polymerization of furfuryl alcohol with ammonium dihydrogen phosphate in poplar wood for improved dimensional stability and flame retardancy. *ACS Sustain. Chem. Eng.* **2018**, *6*, 3349–3357. [CrossRef]
23. Liu, Q.; Chai, Y.; Ni, L.; Lyu, W. Flame retardant properties and thermal decomposition kinetics of wood treated with boric acid modified silica sol. *Materials* **2020**, *13*, 4478. [CrossRef]

24. Kuai, B.; Wang, Z.; Gao, J.; Tong, J.; Zhan, T.; Zhang, Y.; Lu, J.; Cai, L. Development of densified wood with high strength and excellent dimensional stability by impregnating delignified poplar by sodium silicate. *Constr. Build. Mater.* **2022**, *344*, 128282. [CrossRef]
25. Cheng, X.; Lu, D.; Yue, K.; Lu, W.; Zhang, Z. Fire resistance improvement of fast-growing poplar wood based on combined modification using resin impregnation and compression. *Polymers* **2022**, *14*, 3574. [CrossRef]
26. As, N.; Akbulut, T. Odunun fiziksel özelliklerini iyileştiren işlemler ve mekanik özellikler üzerine olan etkisi. *J. Fac. For. Istanb. Univ.* **1989**, *39*, 98–112.
27. LeVan, S.L.; Winandy, E.J. Effect of fire retardant treatment on wood strenght: A Rewiev. *Wood Fiber Sci.* **1990**, *22*, 113–131.
28. Ayrılmış, N. Effects of Various Fire Retardants on Fire and Technological Properties of Some Wood Based Panel Products. Ph.D. Thesis, Istanbul University, Institute of Science, İstanbul, Türkiye, 2006.
29. Demir, A. The Effects of Fire Retardant Chemicals on Thermal Conductivity of Plywood Produced from Different Wood Species. Master's Thesis, Karadeniz Technical University, Institute of Science, Trabzon, Türkiye, 2014.
30. Gökmen, K. The Effect of Heat Treatment with Tall Oil Impregnation on the Properties of Wood Material. Master's Thesis, Bartın University, Institute of Science and Technology, Bartın, Türkiye, 2017.
31. Wang, Y.; Wang, T. Effect of vacuum impregnation on mechanical properties of fast-growing poplar. *J. Northeast For. Univ.* **2019**, *47*, 53–56.
32. Cao, S.; Cai, J.; Wu, M.; Zhou, N.; Huang, Z.; Cai, L.; Zhang, Y. Surface properties of poplar wood after heat treatment, resin impregnation, or both modifications. *BioResources* **2021**, *16*, 7562–7577. [CrossRef]
33. Yang, H.; Gao, M.; Wang, J.; Mu, H.; Qi, D. Fast Preparation of high-performance wood materials assisted by ultrasonic and vacuum impregnation. *Forests* **2021**, *12*, 567. [CrossRef]
34. Zhang, Y.; Guan, P.; Zuo, Y.; Li, P.; Bi, X.; Li, X. Preparation of highly-densified modified poplar wood by evacuating the micro-pores of wood through a gas expansion method. *Ind. Crops Prod.* **2023**, *194*, 116374. [CrossRef]
35. Kausar, A.; Rafique, I.; Anwar, Z.; Muhammad, B. Recent developments in different types of flame retardants and effect on fire retardancy of epoxy composite. *Polym. Plast. Technol. Eng.* **2016**, *55*, 1512–1535. [CrossRef]
36. Blanchet, P.; Pepin, S. Trends in chemical wood surface improvements and modifications: A review of the last five years. *Coatings* **2021**, *11*, 1514. [CrossRef]
37. Kawalerczyk, J.; Walkiewicz, J.; Dziurka, D.; Mirski, R. Nanomaterials to Improve Fire Properties in Wood and Wood-Based Composite Panels. In *Emerging Nanomaterials: Opportunities and Challenges in Forestry Sectors*; Taghiyari, H., Morrell, J.J., Husen, A., Eds.; Springer International Publishing: Cham, Switzerland, 2022; pp. 65–96. [CrossRef]
38. Tan, H.; Şirin, M.; Baltaş, H. Ecological structure: Production of organic impregnation material from mussel shell and combustion. *Polímeros* **2022**, *32*, 1–8. [CrossRef]
39. Zumdahl, S.S.; DeCoste, D.J. *Chemical Principles*, 7th ed.; Cengage Learning: Belmont, MA, USA, 2012.
40. Haynes, W.M. *CRC Handbook of Chemistry and Physics*, 95th ed.; CRC Press: New York, NY, USA, 2014.
41. TS 2470; Methods and General Properties of Sampling from Wood for Physical and Mechanical Experiments. TSE: Ankara, Türkiye, 1976.
42. Kılıç, Ö.; Anıl, M. The effects of limestone characteristic properties and calcination temperature to the lime quality. *Asian J. Chem.* **2006**, *18*, 655–666.
43. Ropp, R.C. *Encyclopedia of the Alkaline Earth Compounds*; Elsevier: Amsterdam, The Netherlands, 2012.
44. Duchesne, J.; Reardon, E.J. Measurement and prediction of portlandite solubility in alkali solutions. *Cem. Concr. Res.* **1995**, *25*, 1043–1053. [CrossRef]
45. ASTM D 1413-99; Standard Method of Testing Wood Preservatives by Laboratory Soil Block Cultures. Annual Book of ASTM Standards: West Conshohocken, PA, USA, 1995.
46. TS EN 2472; Wood—Determination of Density for Physical and Mechanical Tests. Turkish Standards Institution: Ankara, Türkiye, 1972.
47. TS 4084; Wood—Determination of Radial and Tangential Swelling. Turkish Standard Institution: Ankara, Türkiye, 1983.
48. TS EN 317; Particleboards and Fibreboards—Determination of Swelling in Thickness after Immersion in Water. TSE: Ankara, Türkiye, 1999.
49. DIN 4768; Determination of Values of Surface Roughness Parameters, Ra, Rz, Rmax, Using Electrical Contact (Stylus) Instruments. Concepts and Measuring Conditions. Deutsches Institut für Norming: Berlin, Germany, 1990.
50. ASTM D 2863; Standard Test Method for Measuring the Minimum Oxygen Concentration to Support Candle-Like Combustion of Plastics (Oxygen Index). ASTM International: West Conshohocken, PA, USA, 2006.
51. Habizbade, S.; Taghiyari, H.R.; Omidvar, A.; Roudi, H.R. Effects of impregnation with styrene and nano-zinc oxide on fire-retarding, physical, and mechanical properties of poplar wood. *Cerne* **2016**, *22*, 465–474. [CrossRef]
52. Chen, C.; Chen, J.; Zhang, S.; Cao, J.; Wang, W. Forming textured hydrophobic surface coatings via mixed wax emulsion impregnation and drying of poplar wood. *Wood Sci. Technol.* **2020**, *54*, 421–439. [CrossRef]
53. Holy, S.; Temiz, A.; Köse Demirel, G.; Aslan, M.; Amini, M.H.M. Physical properties, thermal and fungal resistance of Scots pine wood treated with nano-clay and several metal-oxides nanoparticles. *Wood Mater. Sci. Eng.* **2022**, *17*, 176–185. [CrossRef]
54. Sogutlu, C.; Dongel, N. The effect of the impregnate process of wooden material to color changes and surface roughness. *J. Polytech.* **2009**, *12*, 179–184.

55. Keskin, H.; Bülbül, R. Impacts of impregnation with Tanalith-E on surface roughness of solid wood materials. *Furnit. Wooden Mater. Res. J.* **2019**, *2*, 67–78. [CrossRef]
56. Aykaç, S.; Sofuoğlu, S.D. A study on the comparison of surface roughness parameters in bamboo material applied with cellulosic, synthetic, polyurethane and water-based varnishes. *Furnit. Wooden Mater. Res. J.* **2020**, *3*, 84–92. [CrossRef]
57. Kartal, S. Wettebality, water absorption and thickness swelling of particleboard made from remediated CCA-treated wood. *J. Fac. For. Istanb. Univ.* **2001**, *51*, 53–62.
58. Kamal, M.R.; Calderon, J.U.; Lennox, B.R. Surface energy of modified nanoclays and its effect on polymer/clay nanocomposites. *J. Adhes. Sci. Technol.* **2009**, *23*, 663–688. [CrossRef]
59. Zaidi, S.J.; Fadhillah, F.; Saleem, H.; Hawari, A.; Benamor, A. Organically modified nanoclay filled thin-film nanocomposite membranes for reverse osmosis application. *Materials* **2019**, *12*, 3803. [CrossRef] [PubMed]
60. Emampour, M.; Khademieslam, H.; Faezipour, M.M.; Talaeipour, M. Effects of coating *Populus nigra* wood with nanoclay. *Bioresources* **2020**, *15*, 8026. [CrossRef]
61. Alhuthali, A.; Low, I.M.; Dong, C. Characterisation of the water absorption, mechanical and thermal properties of recycled cellulose fibre reinforced vinyl-ester eco-nanocomposites. *Compos. Part B Eng.* **2022**, *43*, 2772–2781. [CrossRef]
62. Mandal, M.; Maji, T.K. Comparative study on the properties of wood polymer composites based on different modified soybean oils. *J. Wood Chem. Technol.* **2017**, *37*, 124–135. [CrossRef]
63. Kaya, A.I. Fire performance of thermally modified wood impregnated with clay nanomaterials. *Feb. Fresenius Environ. Bull.* **2022**, *31*, 5292–5296.
64. Janotka, I.; Madejova, J.; Števula, L.; Frt'Alová, D.M. Behaviour of Ca(OH)$_2$ in the presence of the set styrene-acrylate dispersion. *Cem. Concr. Res.* **1996**, *26*, 1727–1735. [CrossRef]
65. Bodirlau, R.; Teaca, C.A. Fourier transform infrared spectroscopy and thermal analysis of lignocellulose fillers treated with organic anhydrides. *Rom. J. Phys.* **2009**, *54*, 93–104.
66. Esteves, B.; Velez Marques, A.; Domingos, I.; Pereira, H. Chemical changes of heat-treated pine and eucalypt wood monitored by FTIR. *Maderas Cienc. Y Tecnol.* **2013**, *15*, 245–258. [CrossRef]
67. Wada, K. A structural scheme of soil allophane. *Am. Mineral.* **1967**, *52*, 690–708.
68. Bellamy, L.J. *The Infra-Red Spectra of Complex Molecules*; John Wiley & Sons: New York, NY, USA, 1966.
69. Wang, X.; Romero, M.Q.; Zhang, X.Q.; Wang, R.; Wang, D.Y. Intumescent multilayer hybrid coating for flame retardant cotton fabrics based on layer-by-layer assembly and sol–gel process, RSC Adv. 2015, 5, 10647–10655. *RSC Adv.* **2015**, *5*, 10647–10655. [CrossRef]
70. Beram, A.; Yaşar, S. Performance of brutian pine (*Pinus brutia* Ten.) particles modified with NaOH in board production. *J. Grad. Sch. Nat. Appl. Sci. Mehmet Akif Ersoy Univ.* **2018**, *9*, 187–196. [CrossRef]
71. Faix, O.; Meier, D.; Fortmann, I. Thermal degradation products of wood. A collection of electron-impact (EI) mass spectra of monomeric lignin derived products. *Holz. Als. Roh. Werkst.* **1990**, *48*, 351–354. [CrossRef]
72. Kotilainen, R.; Toivannen, T.; Alén, R. FTIR monitoring of chemical changes in softwood during heating. *J. Wood Chem. Technol.* **2000**, *20*, 307–320. [CrossRef]
73. Windeisen, E.; Strobel, C.; Wegener, G. Chemical changes during the production of thermotreated beech wood. *Wood Sci. Technol.* **2007**, *41*, 523–536. [CrossRef]
74. Moser, F.; Trautz, M.; Beger, A.L.; Löwer, M.; Jacobs, G.; Hillringhaus, F.; Wormit, A.; Usadel, B.; Reimer, J. Fungal mycelium as a building material. In Proceedings of the Annual Symposium of the International Associationfor Shell and Spatial Structures, IASS 2017, Hamburg, Germany, 25–28 September 2017.
75. Xu, F.; Zhong, L.; Zhang, C.; Wang, P.; Zhang, F.; Zhang, G. Novel high-efficiency casein-based P–N-containing flame retardants with multiple reactive groups for cotton Fabrics. *ACS Sustain. Chem. Eng.* **2019**, *7*, 13999–14008. [CrossRef]
76. Abidin, Z.; Matsue, N.; Henmi, T. Nanometer-scale chemical modification of nanoball allophane. *Clays Clay Miner.* **2007**, *55*, 443–449. [CrossRef]
77. Jiang, J.; Li, J.; Hu, J.; Fan, D. Effect of nitrogen phosphorus flame retardants on thermal degradation of wood. *Constr. Build. Mater.* **2010**, *24*, 2633–2637. [CrossRef]
78. Lowden, L.A.; Hull, T.R. Flammability behaviour of wood and a review of the methods for its reduction. *Fire Sci. Rev.* **2013**, *2*, 1–19. [CrossRef]
79. Zhang, L.L.; Xu, J.S.; Shen, H.Y.; Xu, J.Q.; Cao, J.Z. Montmorillonite-catalyzed furfurylated wood for flame retardancy. *Fire Saf. J.* **2021**, *121*, 103297. [CrossRef]
80. Ghosh, P.; Siddhanta, S.K.; Chakrabarti, A. Characterization of poly (vinyl pyrrolidone) modified polyaniline prepared in stable aqueous medium. *Eur. Polym. J.* **1999**, *35*, 699–710. [CrossRef]
81. Sun, J.X.; Xu, F.; Sun, X.F.; Xiao, B.; Sun, R.C. Physico-chemical and thermal characterization of cellulose from barley straw. *Polym. Degrad. Stab.* **2005**, *88*, 521–531. [CrossRef]
82. Vazquez, A.; Foresti, M.L.; Cerrutti, P.; Galvagno, M. Bacterial cellulose from simple and low cost production media by Gluconacetobacter xylinus. *J. Polym. Environ.* **2013**, *21*, 545–554. [CrossRef]
83. Kozakiewicz, P.; Drożdżek, M.; Laskowska, A.; Grześkiewicz, M.; Bytner, O.; Radomski, A.; Zawadzki, J. Effects of thermal modification on selected physical properties of sapwood and heartwood of black poplar (*Populus nigra* L.). *BioResources* **2019**, *14*, 8391–8404. [CrossRef]

84. Yaşar, S.; Güler, G. Chemical characterization of black poplar (*Populus nigra* L.) sawdust hemicelluloses esterified with acyl chlorides. *Turk. J. For.* **2021**, *22*, 426–431. [CrossRef]
85. Aydemir, D.; Çivi, B.; Alsan, M.; Can, A.; Sivrikaya, H.; Gündüz, G.; Wang, A. Mechanical, morphological and thermal properties of nano-boron nitride treated wood materials. *Maderas. Cienc. Y Tecnol.* **2016**, *18*, 19–32. [CrossRef]

Disclaimer/Publisher's Note: The statements, opinions and data contained in all publications are solely those of the individual author(s) and contributor(s) and not of MDPI and/or the editor(s). MDPI and/or the editor(s) disclaim responsibility for any injury to people or property resulting from any ideas, methods, instructions or products referred to in the content.

Article

Experimental Evaluation of Glulam Made from Portuguese Eucalyptus

Aiuba Suleimana [1,2], Bárbara C. Peixoto [3], Jorge M. Branco [1,*] and Aires Camões [4]

[1] Department of Civil Engineering, University of Minho, ISISE, 4800-058 Guimarães, Portugal; asuleimana@unilurio.ac.mz
[2] Rural Engineering Department, Lúrio University, Sanga 3302, Mozambique
[3] School of Architecture, Art and Design, University of Minho, 4800-058 Guimarães, Portugal
[4] Centre for Territory, Environment and Construction (CTAC), University of Minho, 4800-058 Guimarães, Portugal; aires@civil.uminho.pt
* Correspondence: jbranco@civil.uminho.pt

Abstract: Engineered wood products (EWPs) have evolved over time to become a popular and sustainable alternative to traditional lumber by offering design flexibility, increased strength, and improved quality control. This work analyzes the potential of Portuguese eucalyptus wood (*Eucalyptus globulus*) to produce glued-laminated timber (glulam) for structural applications. Currently, this hardwood is used for less noble applications in Portugal's construction industry. To promote the use of this species of timber in construction, an experimental campaign was conducted to characterize its compression parallel to the grain and bending strength. The results demonstrated that this hardwood presents a compression parallel to the grain strength of 73 N/mm^2 and a bending strength of 151 N/mm^2 with a global value of elastic modulus equal to 24,180 N/mm^2. Based on those strength values obtained from the glulam produced with eucalyptus, one can conclude that the test results presented here are higher than the ones declared by the current glulam made of softwoods; thus, additional studies are encouraged.

Keywords: glulam; hardwoods; eucalyptus; experimental evaluation

Citation: Suleimana, A.; Peixoto, B.C.; Branco, J.M.; Camões, A. Experimental Evaluation of Glulam Made from Portuguese Eucalyptus. *Appl. Sci.* **2023**, *13*, 6866. https://doi.org/10.3390/app13126866

Academic Editors: Laurent Daudeville, Claudio De Pasquale, Alena Očkajová, Martin Kučerka and Richard Kminiak

Received: 26 April 2023
Revised: 27 May 2023
Accepted: 31 May 2023
Published: 6 June 2023

Copyright: © 2023 by the authors. Licensee MDPI, Basel, Switzerland. This article is an open access article distributed under the terms and conditions of the Creative Commons Attribution (CC BY) license (https:// creativecommons.org/licenses/by/ 4.0/).

1. Introduction

The Portuguese Institute of Nature Conservation and Forests (ICNF) shows that the *Eucalyptus globulus* occupies 26.2% of all existing forests in the country, making it, already, the most representative wood species in Portugal [1]. This hardwood is originally from Australia and was introduced in Portugal in 1954 [2]. However, due to the fast-growing nature of the tree, with short harvesting times between the ages of 10 and 15 years, compared to the most representative local species, such as maritime pine (35 to 45 years) [3], in addition to its easy adaptation to soil and atmospheric conditions, made its introduction in the Portuguese forest easy and fast. Moreover, the growth of the national paper industry made it economically interesting for private proprietors to plant eucalyptus.

In Portugal, eucalyptus is primarily used for pulp and paper production, as well as for biomass as a source of energy [4]. In construction, this species has been used for parquet flooring for residential buildings, and for traditional applications, such as railway sleepers, mine structures [5], and for sports purposes, due to its hardness and clear color. In the past, in particular in rural regions, as a consequence of its availability and low cost, it has been used for structures such as roofs, floors, and walls. For example, the historical center of Guimarães is well known for its half-frame timber construction system of using eucalyptus wood since the second half of the 20th century.

The production of glued-laminated timber (glulam) normally uses softwoods such as spruce, fir, and Scots pine, which are the most common species in Central Europe and are where the most important manufacturers of glulam are located. By contrast, the

Portuguese forests present a high percentage of hardwoods, with eucalyptus being the most representative species. In this context, it will be interesting to study the possibility of using eucalyptus to produce glulam, thereby adding value to the Portuguese forests. It is possible to find German producers of glulam, which is made of beech from the black forest; however, to our knowledge, the only producers of glulam made from eucalyptus are located in Galicia, Spain, where the presence of this species is also important. Eucalyptus seeds were first introduced in Spain in the 1800s [6], and their woods have been used in different areas, from mercenaries to carpentry and buildings.

Most of the glulam is produced using softwoods; however, the use of some hardwoods was previously mentioned in EN 14080 [7]. Unfortunately, *Eucalyptus globulus* was not mentioned by this standard as a possible species to use in the production of glulam.

Seng Hua et al. [8] presented a summary of studies conducted on the different uses of eucalyptus species worldwide, with special emphasis on engineered wood products (EWP), such as glulam, cross-laminated timber (CLT), fiberboards, oriented strand boards (OSBs), laminated veneer timber (LVT), and other particleboards and plywood. It was demonstrated that glulam was the application least studied.

In Brazil, glulam beams made from *Eucalyptus urogandis* have been tested and obtained values for bending strength in the range of 57–94 N/mm^2, with an elastic modulus in the range of 17,800–20,100 N/mm^2, a compression strength parallel to the grain between 65 and 75 N/mm^2, and a bond line shear strength of 10–12 N/mm^2 [9], which suggested that the species had great potential to be applied in glulam industries owing to its high mechanical properties.

Despite this lack of information, Franke and Marto [10] evaluated the use of *Eucalyptus globulus* as an engineered material, achieving characteristic values of 115 N/mm^2 and 83 N/mm^2 for bending and compressive strength parallel to the grain, respectively. Moreover, the gluing and delamination properties were measured with PUR adhesive, and it was observed that none of the tried configurations were successful in meeting all the requirements regarding the strength and durability of the glue lines. Nevertheless, glue lines based on MUF adhesive were studied by Suleimana et al. [11], who obtained failures that occurred mostly on the wood side and achieved shear strengths of 14 N/mm^2 and 12 N/mm^2 using surface-bonding and edge-bonding specimens, respectively.

In Spain, solid and glulam elements made of *Eucalyptus globulus* have been evaluated and characterized by Lara-Bocanegra et al. [12], for use in timber grid shells. A bending strength of 55 N/mm^2 was achieved by the solid laths, while a strength class higher than a GL56 could be attributed to the glulam elements. In Portugal, homogeneous glulam beams made of *Eucalyptus globulus* and hybrid (eucalyptus and poplar) were evaluated by Martins et al. [13]. The average values obtained for bending modulus were 23,487–25,615 N/mm^2 and 18,302–22,341 N/mm^2 for local and global values, respectively, while the bending strength ranged from 91 to 115 N/mm^2 for the hybrid and homogenous glulam beams, respectively.

In the current work, an experimental evaluation of glulam beams made from *Eucalyptus globulus* is presented. Glulam beams were manufactured in an industrial facility and tested in the laboratory under four-point bending tests. Due to the lack of data about the mechanical characterization of the eucalyptus wood, small specimens, representing the lamellae used in the glulam beams, were used to prepare compression parallel to the grain and bending tests. The objective was to assess the mechanical characterization of the raw material used in the lamellae. The National Laboratory of Civil Engineering (LNEC) published results of an experimental campaign based on small clear samples of *Eucalyptus globulus* [14], which were used as references. Density values of 750–850 kg/m^3, bending strength of 128 N/mm^2, and a modulus of elasticity in bending of 17,500 N/mm^2 were demonstrated, while compression and tension strengths parallel to the grain were equal to 49 N/mm^2 and 14 N/mm^2, respectively, alongside a shear strength of 3 N/mm^2.

The main objective of the present work was to assess the possibility of producing glulam from *Eucalyptus globulus* grown in Portugal, to add value to this hardwood species, while reducing the national dependence on imported softwoods.

2. Materials and Methods

The eucalyptus specimens were collected from logs cut in north Portugal by a sawmill located close to Amarante. After their preparation and transport to laboratories at the University of Minho, all specimens were kept in a climatic chamber under a controlled temperature of 20 °C and relative humidity (RH) of 60% for approximately 4 weeks, until mass stabilization was reached, as recommended by NP 614 and ISO 3130 [15,16]. The lamellae for the glulam beam production were prepared in the *Rusticasa* industry facilities and the tests were conducted in the Laboratories of the Civil Engineering Department at the University of Minho. Table 1 summarizes the experimental campaign performed with the identified corresponding standards as follows.

Table 1. Summary of the experimental campaign performed.

Property	Symbol	Unity	Standard
Moisture content	ω	%	ISO 3130 [16]
Density	ρ_k	kg/m^3	EN 13183-1 [17]
Compression parallel to the grain (clear specimens)	$f_{c,0}$	N/mm^2	ASTM D143 [18]
Bending strength (clear specimens)	$f_{m,90}$	N/mm^2	ASTM D143 [18]
Bending strength on structural beams	$f_{m,90}$	N/mm^2	EN 408 [19]

2.1. Determination of Moisture Content and Density

After the stabilization of the specimens inside the climatic chamber, they were removed, and the dimensions and weight were measured to determine the density (ρ). Afterward, the tested specimens were conditioned in an oven to dry and the moisture content (ω) was measured. Both characteristics of the specimens were determined by equations that can be found in ISO 3130 and EN 13183-1, respectively [16,17].

2.2. Compression Parallel to the Grain on Clear Specimens

The purpose of this test was to evaluate the compression strength parallel to the grain in clear specimens by following the guidelines in ASTM D143 [18]; the specimens were produced by following NP 618 [20]. These results allowed a control quality to be performed on the raw material of *Eucalyptus globulus* lamellae. Moreover, it was, then, possible to compare the results to the values suggested by LNEC [14].

The specimen was placed on the bottom plate of the machine centered on the vertical axis, as shown in Figure 1a,b. The actuator was lowered until it touches the specimen's face. Then, the load, registered by a load cell with a maximum capacity of 200 kN, was applied at a rate of 0.01 mm/second until the specimen was broken, which normally happened after 3 (+/−2) min.

To determine the results of compression strength parallel to the grain, Equation (1) was applied.

$$f_{c,0} = \frac{F_{max}}{bh} \qquad (1)$$

where $f_{c,0}$, is the compression strength parallel to the grain in N/mm^2, F_{max} is the maximum load in Newton, and b and h are the width and height of the specimen cross-section, respectively, measured in millimeters. In total, fourteen clear specimens were tested.

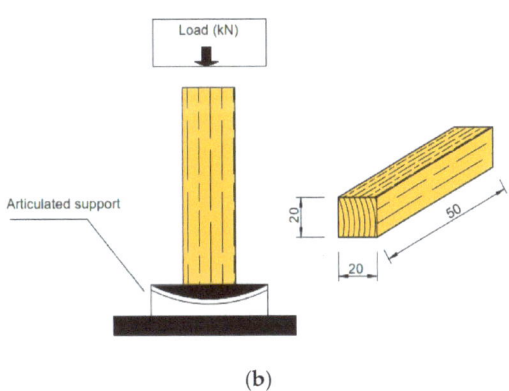

Figure 1. Compression parallel to the grain. (**a**) Test setup, (**b**) specimen.

2.3. Bending Strength on Clear Specimens

For the bending strength in clear specimens, tests according to NP 619 [21] and ASTM D143 [18] were performed. The machine's actuator was lowered until it touched the specimen's face. Then, a maximum actuator force of 25 kN with a constant speed of 0.01 mm/second was applied to ensure that the rupture occurred over a maximum period of 3 (+/−2) min. Figure 2a,b shows a representative bending failure and the setup used.

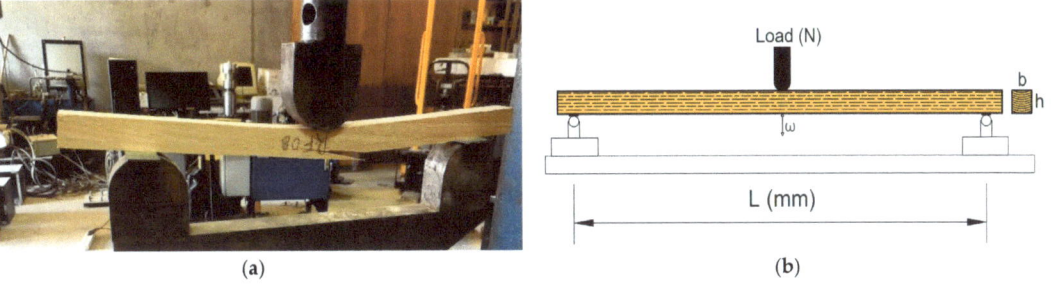

Figure 2. Configuration of four-point bending test according to NP 619 [21] and ASTM D143 [18]; (**a**) bending failure, (**b**) setup.

To determine the results of bending strength on the clear specimens, Equation (2) was applied, whereas to obtain the values of the moisture content and density, measurements for the weight and volume methods already described were applied to the specimens after testing the bending.

$$f_m = \frac{3Fl}{2bh^2} \left[\text{N/mm}^2 \right] \qquad (2)$$

where f_m is the static bending tensile strength, expressed in N/mm^2, F is the maximum bending force recorded during the test, in Newton, l is the span between the beam supports in millimeters, and b and h are the width and height of the specimen cross-section, respectively, in millimeter. In total, fifteen clear specimens were tested.

2.4. Production of the Glulam Beams

For the gluing of the glulam beams, a two-component of glue, so-called melamine-urea-formaldehyde MUF 1242/2542, produced by Akzo Nobel, was used (Table 2).

Table 2. MUF 1247/2526 technical specifications [22].

Component	1247			2526		
Product	MUF glue			Hardener		
Appearance	Liquid			Liquid		
Color	Opaque/White			White		
Viscosity (At production moment)	10,000–25,000 mPas (Brookfield LVT, sp.4, 12 rpm, 25 °C/77 °F)			1700–2700 mPas (Brookfield LVT, sp.4, 60 rpm, 25 °C/77 °F)		
PH (At production moment)	9.5–10.7 (at 25 °C/77 °F)			1.3–2.0 (at 25 °C/77 °F)		
Dry layer	64–69%			Not applicable		
Storage time (months)	15 °C/59 °F 4	20 °C/68 °F 4	30 °C/86 °F 2	15 °C/59 °F 4	20 °C/68 °F 4	30 °C/86 °F 2.5
Information on formaldehyde	Free formaldehyde ≤ 0.8%			Formaldehyde free		
Density	Approx. 1270 kg/m^3			Approx. 1070 kg/m^3		
Properties of the glue	Lightly colored glue joints. High resistance to water and adverse environmental conditions. Meets the requirements of EN 301 [23] (for glue type I and II, service class 1, 2, and 3), EN 391 [24], EN 392 [25], and DIN 68141 [26].					

The production process began by sorting the boards and defining the beam configuration. The process of beam manufacturing was conducted by the *Rusticasa* Company as described below.

Firstly, the eucalyptus wood was obtained, and then, cut to produce lamellas of 2500 mm in length and with a transversal cross-section of 76 × 21 mm^2. After that, the lamellas were classified visually to make a control and remove the most defective lamellas; the approved ones are shown in Figure 1a. Then, the approved lamellas were kept in a climatic chamber for three months at a temperature of 20 °C and a relative humidity of 60%, for the glue curing to occur. Periodic measurement was performed to control moisture stabilization. After drying, the lamellas were removed from the climatic chamber and planed on the same day as the glue was applied (less than 6 h).

With the glue applicator on, as shown in Figure 3b, one person holds the first lamellae and passes it from the middle of the glue applicator to another person and the second one applies it to the press bed. This process provides a possibility for the uniform application of the glue and was performed for seven lamellae before it was interrupted. To separate between beams, the lamellae, without glue, were placed on the top of the group and the same process of passing the lamellas from the middle of the glue applicator was repeated. This step required 20 min to be completed for all the beams.

After all the lamellas were glued and placed on the press bed, they were subjected to a pressure of between 133 bar and 135 bar, for one hour, for the gluing and curing to occur, as shown in Figure 1c. Thereafter, they were removed from the press, the remaining glue was cleaned and the faces and the corners were adjusted for the dimensions, shown in Figure 1d. Finally, they were taken to the Laboratories of the Civil Engineering Department of the University of Minho for testing. Note that, the same procedure mentioned above was followed by Martins et al. [27] and Pedro et al. [28] in the production of glulam using other hardwoods (*Acacia melanoxylon* and *Acacia mangium*, respectively).

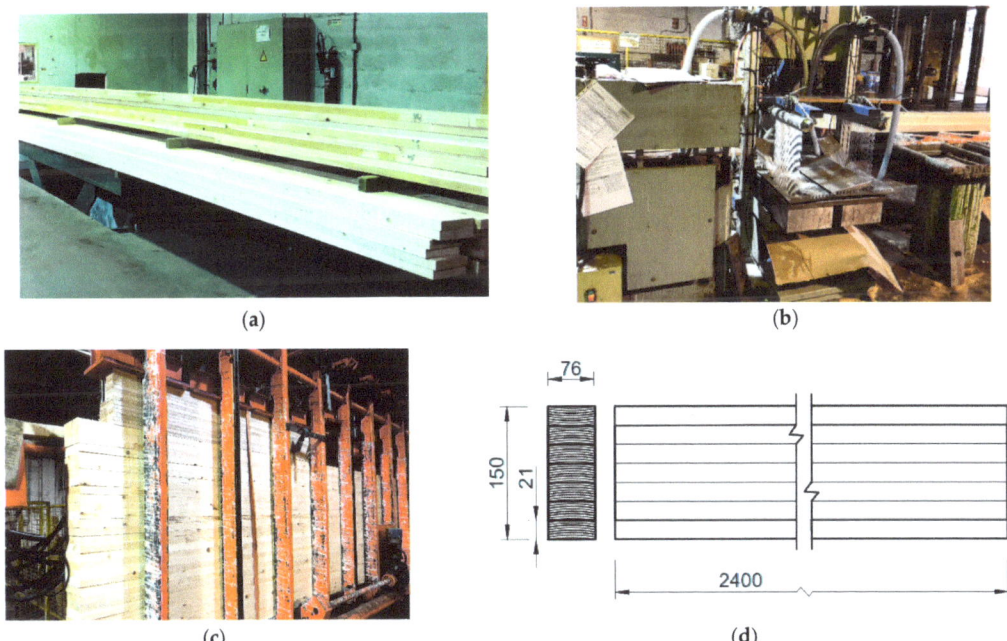

Figure 3. Different stages of the industrial process to produce glulam and the configuration of beams; (**a**) paired boards ready for glue application, (**b**) glue machine in use, (**c**) boards with glue stacked and applying pressure for gluing, (**d**) configuration of beams in millimeters.

2.5. Bending Tests on Beams

The bending tests performed on structural glulam beams produced with eucalyptus aimed to quantify their main bending properties, namely, global values of the modulus of elasticity in bending ($E_{m,g}$) and the corresponding bending strength (f_m). These mechanical properties were measured following the recommendation made by EN 408 [19] with regard to the specimen's geometry, test, setup, loading protocol, and instrumentation. Figure 4 illustrates the four-point bending test on a glulam beam made from eucalyptus. In total, four glulam beams were tested.

Figure 4. Four-point bending test on a glulam beam.

Bending strength of the beams is determined by Equation (3):

$$f_m = \frac{3aF_{max}}{bh^2} \ [\text{N/mm}^2] \tag{3}$$

where F_{max} is the maximum load applied in Newton, b and h are the width and height of the beam's cross-section, respectively, in millimeters, and a is the distance between the point where the load is applied and its closest support.

In addition, the global value of the elastic modulus in bending is obtained by Equation (4).

$$E_{m,g} = \frac{l^3(F_2 - F_1)}{bh^3(w_2 - w_1)} \left[\left(\frac{3a}{4l}\right) - \left(\frac{a}{l}\right)^3 \right] \ [\text{N/mm}^2] \tag{4}$$

where the F_1 and F_2 are the loadings corresponding to 10% and 40% of the maximum load, respectively, in Newton, w_1 and w_2 are the displacements corresponding to loads of F_1 and F_2, respectively, in millimeters, l is the distance between the two supports in millimeters, and b, h and a, again, assume the meaning already presented for Equation (3).

3. Results

Here, all test results are presented and, when possible, comparisons with existing values are made.

3.1. Results of Clear Specimens

Density (ρ) and moisture content (ω) of all specimens, as already mentioned, were quantified according to ISO 3130 [16] and EN 13183-1 [17], respectively. In the first phase, the density was calculated considering the boards with their initial dimensions before the bending tests. Subsequently, the density was measured after the mechanical tests near the failure zones for both structural and smaller specimens. The results obtained are summarized in Table 3. The results of the mechanical tests performed on small clear specimens for bending and compression, parallel to the grain, are depicted in Figure 5. The failures obtained were, as expected, tension and crushing, respectively, which are presented in Figure 5c,d.

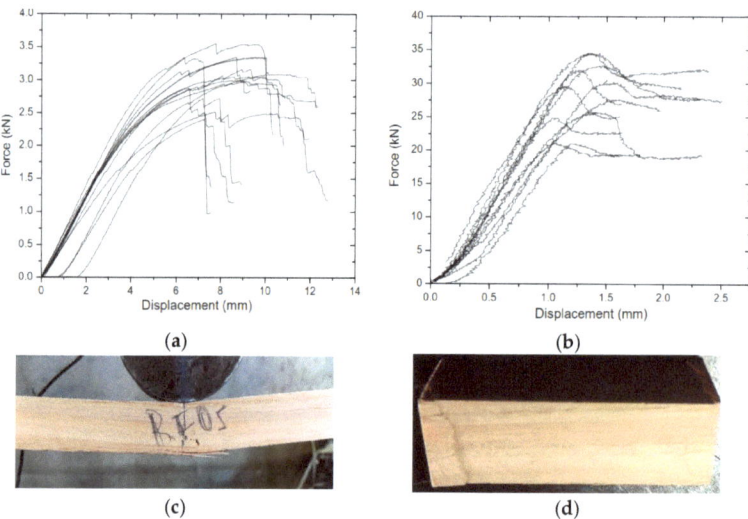

Figure 5. Experimental load-displacement curves obtained from the bending (**a**) and compression parallel to the grain (**b**) tests, and respective failure mode associated, tension (**c**) under bending and crush (**d**) under compression parallel to the grain.

Table 3. Results of physical and mechanical characterization on clear specimens.

Characteristic		ρ (kg/m³)	ω (%)	$f_{c,0}$ (N/mm²)	f_m (N/mm²)
Compression parallel to the grain	Mean	802.43	9.65	72.84	
	Maximum	845.42	10.87	86.35	
	Minimum	691.53	8.74	52.5	
	CoV (%)	8.51	6.79	13.25%	
Bending	Mean	809.5	8.86		151.34
	Maximum	958.21	10.19		175.88
	Minimum	615	6.58		118.65
	CoV (%)	12.33	9.24		11.66%

Table 3 presents the average results obtained from the compression parallel to the grain and bending tests, corresponding to the 14 and 15 tested specimens, respectively, as well as its maximum, minimum, and coefficient of variation, CoV (%).

The results obtained for the clear specimens show similar values of strength for bending (f_m) and compression, parallel to the grain ($f_{c,0}$), which were calculated according to NP 618 [20] and NP 619 [21], respectively, and were also conciliated and supported by ASTM D143 [18], which correspond to averages of 72.84 N/mm² and 151.34 N/mm², respectively. The coefficient of variation was minus or equal to 13.26%. It is important to note that the observed failure on the clear specimens was because of tension under the bending strength (Figure 5c) and by crush under compression strength, parallel to the grain (Figure 5d). Moreover, the density ranged between 802 kg/m³ and 813 kg/m³, with a CoV of 12.5%, while the moisture content varied between 9% and 11%.

3.2. Bending Tests on Glulam Beams

The bending test results obtained for glulam beams produced in this study with Portuguese eucalyptus are presented in Table 4.

Table 4. Results of the bending tests on the glulam beams produced from eucalyptus.

	ρ (kg/m³)	ω (%)	f_m (N/mm²)	E_{mg} (N/mm²)
Mean	812.70	11.07	96.74	24,180
Maximum	854.83	11.65	106.16	26,300
Minimum	777.19	10.79	89.36	22,580
CoV (%)	3.68	3.13	7.37	6.46

The glulam beams made from eucalyptus, on average, have a density of 812.70 kg/m³ and a bending strength of 96.74 N/mm², which is a value three times bigger than the higher strength class for glulam made of softwoods, GL32 [7]. Moreover, the results obtained for the modulus of elasticity in bending ranges from 22,580 N/mm² to 26,300 N/mm², which are values approximately double the one of GL32 (32 N/mm² and 14,200 N/mm² for bending strength and global elasticity modulus, respectively). The force versus displacement curves are presented in Figure 6a and it is important to point out that all beam tests have reached failure modes typical for bending parallel to the grain, with a failure by pure tension and progress of the failure by shear (Figure 6b,c).

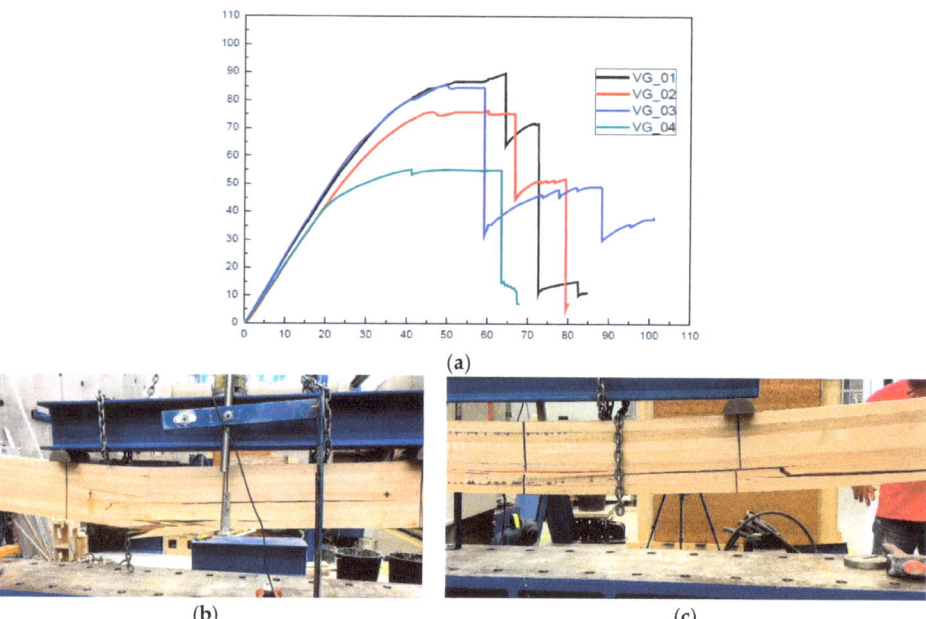

Figure 6. Curves of applied loads versus displacement (**a**). Typical failure modes of timber elements under bending. (**b**) Failure by pure tension and (**c**) failure progress by shear.

4. Discussion

Table 5 compares the results reported by the National Laboratory of Civil Engineering (LNEC) based on small clear specimens of *Eucalyptus globulus* and the average values obtained in the experimental campaign performed.

Table 5. Summary of the properties evaluated and reported by LNEC for the Portuguese *Eucalyptus globulus*.

Properties		Unity	LNEC [14]	Obtained Results
Density		kg/m^3	750–850	803–813
Bending strength	Module of Rupture	N/mm^2	127.5	151.3
Compression parallel to the grain	Strength of failure	N/mm^2	49.1	72.8
Bending strength (Glulam beams)	Module of Rupture Elasticity modulus	N/mm^2	- -	96.7 24,180

The specimens measured have an average density of 805 kg/m^3, corresponding to a moisture content of clear and small-sized, which was taken from glulam specimens less or equal to 12.5%, as recommended by EN 13183-1 [17]. These values are similar to the corresponding results from LNEC [14], although they are less than the one presented by Franke, and Marto [10] (980 kg/m^3) and higher than the one presented by Almeida et al. [29] (730 kg/m^3). Therefore, compared to results from LNEC and previous works, the values obtained in this experimental campaign can be assumed to be coherent and slightly better.

The compression parallel to the grain and bending strength attained average values of 73 N/mm^2 and 151 N/mm^2, respectively. These values are higher than the ones presented by LNEC [14], by 48% and 19%, respectively. Regarding the ones presented by Franke and Marto [10], similar values were achieved for the compression strength parallel to the grain

(83 N/mm^2, with a 14% difference) while it was higher than the bending strength value (115 N/mm^2). Moreover, this study pointed out experimental values higher than those presented by Acosta [30] for the compression strength parallel to the grain and bending strength, which were 89 N/mm^2 and 116 N/mm^2, respectively. This variety between the values presented in distinct studies can be justified by different parameters such as the origin and the age of the trees. It is worth highlighting that the average bending strength and global elastic modulus values of 97 N/mm^2 and 24,180 N/mm^2, respectively, registered for the glulam beams tested according to EN 408 [19], were higher than the maximum ones suggested by EN 14080 [7] for glulam elements made of softwoods. Moreover, they were higher than the ones obtained by an experimental campaign, conducted by Nogueira et al. [9], on full-scale glulam beams manufactured with *Eucalyptus urograndis* in Brazil. Nogueira et al. [9] achieved a bending strength in the range of 57–94 N/mm^2 and 17,812–20,072 N/mm^2 for the modulus of elasticity, with a compression strength parallel to the grain strength of 65–75 N/mm^2.

The results obtained in the experimental campaign performed demonstrate the promising mechanical properties of glulam made with *Eucalyptus globulus*, which can be considered a competitive EWP for structural applications and an excellent alternative to the current glulam made of softwoods.

5. Conclusions

The experimental campaign performed, and here presented, contributes to the hypothesis that *Eucalyptus globulus* from Portuguese forests can be used to produce glulam. Firstly, it was demonstrated that it is possible to produce glulam made from Portuguese eucalyptus in an industrial environment without any special provisions or changes to the regular production process, including gluing. Then, the obtained mechanical properties of the glulam made of eucalyptus were shown to be superior to the high strength class defined for current glulam made of softwood (GL32). Therefore, one can conclude that the use of this hardwood species, which is already the dominant species in Portugal, can contribute to adding value to Portuguese forests. It is important to note that the glulam produced with Portuguese *Eucalyptus globulus* exhibited mechanical properties higher than those found in the literature. However, only with a full and detailed characterization of this wood species, and the glulam elements produced, can their use in construction be considered. In particular, the assessment of the presence of finger-joints in the response to the glulam elements must be performed in future research since, in this study, no finger-joints were considered.

Author Contributions: Conceptualization, J.M.B. and A.C.; Methodology, A.S., B.C.P. and A.C.; Validation, J.M.B.; Investigation, A.S. and B.C.P.; Resources, J.M.B.; Writing–original draft, A.S.; Writing–review & editing, B.C.P., J.M.B. and A.C.; Supervision, A.C.; Funding acquisition, J.M.B. All authors have read and agreed to the published version of the manuscript.

Funding: This work was financed by Transform Agenda, approved under notice N°02/C05-i01/2022. Investment supported by the PRR-Recovery and Resilience Plan and by the NextGeneration EU European Funds and through a PhD grant SFRH/BD/151442/2021 conceded to the first author.

Institutional Review Board Statement: Not applicable.

Informed Consent Statement: Not applicable.

Data Availability Statement: Not applicable.

Conflicts of Interest: The authors declare no conflict of interest.

References

1. Malico, I.; Gonçalves, A.C. *Eucalyptus globulus* coppices in portugal: Influence of site and percentage of residues collected for energy. *Sustainability* **2021**, *13*, 5775. [CrossRef]
2. Alves, A.M.; Pereira, J.S.; Silva, J.M.N. *The Eucalyptus in Portugal Environmental Impacts and Scientific Research*; ISAPress; Colprinter, Indústria Grafica, Lda: Lisbon, Portugal, 2007. Available online: https://rb.gy/yzqpb (accessed on 30 May 2023).

3. Dias, A.C.; Arroja, L. Environmental impacts of eucalypt and maritime pine wood production in Portugal. *J. Clean. Prod.* **2012**, *37*, 368–376. [CrossRef]
4. Barreiro, S.; Tomé, M. Analysis of the impact of the use of eucalyptus biomass for energy on wood availability for eucalyptus forest in Portugal: A simulation study. *Ecol. Soc.* **2012**, *17*, 14. [CrossRef]
5. Floresta, P.T. "Forestry and Industry," Silvicultura e Industria. 2023. Available online: https://shre.ink/kVyV (accessed on 20 March 2023).
6. Ruiz, F.; López, G. Review of cultivation history and uses of eucalypts in Spain. In Proceedings of the Conference of Eucalyptus Species Management, History, Status and Trends in Ethiopia, Addis Ababa, Ethiopia, 15–17 September 2010.
7. *EN 14080*; BSI Standards Publication Timber structures—Glued Laminated Timber and Glued Solid Timber—Requirements. British Standards Institute: London, UK, 2014; pp. 1–110.
8. Hua, L.S.; Chen, L.W.; Antov, P.; Kristak, L.; Md Tahir, P. Engineering Wood Products from *Eucalyptus* spp. *Adv. Mater. Sci. Eng.* **2022**, *2022*, 8000780. [CrossRef]
9. Nogueira, R.d.S.; Icimoto, F.H.; Calil Junior, C.; Rocco Lahr, F.A. Experimental study on full-scale glulam beams manufactured with Eucalyptus urograndis. *Maderas Cienc. Tecnol.* **2022**, *25*, 1–12. [CrossRef]
10. Franke, S.; Marto, J. Investigation of *Eucalyptus globulus* wood for the use as an engineered material. In Proceedings of the WCTE 2014—World Conference on Timber Engineering, Quebec City, QC, Canada, 10–14 August 2014.
11. Suleimana, A.; Sena, C.S.; Branco, J.M.; Camões, A. Ability to Glue Portuguese Eucalyptus Elements. *Buildings* **2020**, *10*, 133. [CrossRef]
12. Lara-bocanegra, A.J.; Majano-majano, A.; Arriaga, F.; Guaita, M. *Eucalyptus globulus* finger jointed solid timber and glued laminated timber with superior mechanical properties: Characterisation and application in strained gridshells. *Constr. Build. Mater.* **2020**, *265*, 120355. [CrossRef]
13. Martins, C.; Dias, A.M.P.G.; Cruz, H. Blue gum: Assessment of its potential for glued laminated timber beams. *Eur. J. Wood Wood Prod.* **2020**, *78*, 905–913. [CrossRef]
14. LNEC (National Laboratory of Civil Engineering). *Wood for Construction_Eucalyptus Globulus*; National Laboratory for Civil Engineering: Lisboa, Portugal, 2014.
15. *NP 614*; Water Content Detremination. National Laboratory for Civil Engineering: Lisboa, Portugal, 1973; p. 2.
16. *ISO 3130*; Wood-Determination of Moisture Content for Physical and Mechanical Tests. International Organization for Standardization: Geneva, Switzerland, 1975; Volume 1975, p. 4.
17. *EN 13183-1*; Moisture Content of a Piece of Sawn Timber—Determination by Oven Dry Method. European Committee for Standardization CEN: Brussels, Belgium, 2002; Volume 53, pp. 1689–1699.
18. *ASTM D143-94*; Standard Methods of Testing Small Clear Samples of Timber. American Society for Testing and Materials: West Conshohocken, PA, USA, 1994; p. 31.
19. *EN 408*; Structural Timber and Glued Laminated Timber Determination of Some Physical and Mechanical Properties. British Standards Institute: London, UK, 2004; p. 30.
20. *NP 618*; Axial Compression Test. National Laboratory for Civil Engineering: Lisboa, Portugal, 1973; pp. 1–3.
21. *NP 619*; Static Bending Test. National Laboratory for Civil Engineering: Lisboa, Portugal, 1973; pp. 1–3.
22. Nobel, A. Product Information Product Information. Opadry. 2009, p. 6. Available online: https://rb.gy/jgvvu (accessed on 30 May 2023).
23. *EN 301*; Adhesives, Phenolic and Aminoplastic, for Load-Bearing Timber Structures–Classification and Performance Requirements. Comité Européen de Normalisation CEN: Brüssel, Belgium, 2006.
24. *EN 391*; Delamination Test of Glue lines. European Committee for Standardization: Brüssel, Belgium, 2002.
25. *EN 392*; Shear Test of Glue lines. European Committee for Standardization: Brüssel, Belgium, 1995.
26. *DIN 68141:1995-08*; Holzklebstoffe–Prüfung der Gebrauchseigenschaften von Klebstoffen für tragende Holzbauteile. Deutsches Institut für Bautechnik: Berlin, Germany, 2008.
27. Martins, C.; Dias, A.M.P.G.; Cruz, H. Bonding performance of Portuguese Maritime pine glued laminated timber. *Constr. Build. Mater.* **2019**, *223*, 520–529. [CrossRef]
28. De Alcântara Segundinho, P.G.; França, L.C.A.; De Medeiros Neto, P.N.; Gonçalves, F.G.; Da Silva Oliveira, J.T. Glued laminated timber (GLULAM) with Acacia mangium and structural adhesives. *Sci. For. Sci.* **2015**, *43*, 533–540.
29. Nogueira, M.C.d.J.A.; de Almeida, D.H.; de Araujo, V.A.; Vasconcelos, J.S.; Christoforo, A.L.; de Almeida, T.H.; Lahr, F.A.R. Physical and mechanical properties of *Eucalyptus saligna* wood for timber structures. *Ambient. Construído* **2019**, *19*, 233–239. [CrossRef]
30. Acosta, M.S.; Mastrandrea, C.; Lima, J.T. Wood Technologies and Uses of Eucalyptus wood from Fast Frown Plantation for Solid Products. In Proceedings of the 51st International Convention of Society of Wood Science and Technology, Concepción, Chile, 10–12 November 2008; pp. 1–12.

Disclaimer/Publisher's Note: The statements, opinions and data contained in all publications are solely those of the individual author(s) and contributor(s) and not of MDPI and/or the editor(s). MDPI and/or the editor(s) disclaim responsibility for any injury to people or property resulting from any ideas, methods, instructions or products referred to in the content.

Article

The Efficiency of Edge Banding Module in a Mass Customized Line for Wooden Doors Production

Zdzisław Kwidziński [1,2], Luďka Hanincová [3], Eryka Tyma [1], Joanna Bednarz [4], Łukasz Sankiewicz [2], Bartłomiej Knitowski [2], Marta Pędzik [1,5], Jiří Procházka [3] and Tomasz Rogoziński [1,*]

1 Department of Furniture Design, Faculty of Forestry and Wood Technology, Poznań University of Life Sciences, Wojska Polskiego 38/42, 60-627 Poznań, Poland
2 Porta KMI Poland, Szkolna 54, 84-239 Bolszewo, Poland
3 Department of Wood Science and Technology, Faculty of Forestry and Wood Technology, Mendel University in Brno, Zemědělská 3, 61300 Brno, Czech Republic
4 Department of International Business, Faculty of Economics, University of Gdańsk, Armii Krajowej 119/121, 81-824 Sopot, Poland
5 Center of Wood Technology, Łukasiewicz Research Network—Poznan Institute of Technology, 60-654 Poznań, Poland
* Correspondence: tomasz.rogozinski@up.poznan.pl

Citation: Kwidziński, Z.; Hanincová, L.; Tyma, E.; Bednarz, J.; Sankiewicz, Ł.; Knitowski, B.; Pędzik, M.; Procházka, J.; Rogoziński, T. The Efficiency of Edge Banding Module in a Mass Customized Line for Wooden Doors Production. *Appl. Sci.* **2022**, *12*, 12510. https://doi.org/10.3390/app122412510

Academic Editor: Abílio Manuel Pinho de Jesus

Received: 16 November 2022
Accepted: 5 December 2022
Published: 7 December 2022

Publisher's Note: MDPI stays neutral with regard to jurisdictional claims in published maps and institutional affiliations.

Copyright: © 2022 by the authors. Licensee MDPI, Basel, Switzerland. This article is an open access article distributed under the terms and conditions of the Creative Commons Attribution (CC BY) license (https://creativecommons.org/licenses/by/4.0/).

Abstract: The TechnoPORTA technology line is a fully automated smart line ensuring the highest quality and efficiency of production wooden doors. The aim of the study was to experimentally determine the performance of the edge banding module in the TechnoPORTA line on particular working days and to determine the possible influence of organizational and technological factors characterizing the line's operation, which can be defined and determined by analyzing the temporal technological data obtained from the IT systems controlling the line's operation. The research was conducted on the edge banding module, which is crucial to the performance of the entire TechnoPORTA line. During the study, data on door leaf machining were collected such as the mean time of production per one working cycle, mean time of retooling, number of retooling, number of door leaves leaving in a series, and most frequent time of series. The data collected by the IT system controlling the line indicates that this module is flexible and its performance is not related to the control parameters. The results can be used to improve the operation of the module and the replication of the work schedule to subsequent modules of the technological line.

Keywords: door industry; TechnoPORTA line; technological line; sustainable business model; IT systems

1. Introduction

Davis [1], the creator of the concept of mass customization (MC), defined it as reaching a large number of customers, as is the case in mass markets, but also treating them individually, as in individualized markets. The assumption of MC is high individualization while maintaining relatively low costs and mass production efficiency [2]. Today, the term MC is used for strategies connected with high variety, personalization, and flexible production [3]. This results from the acceptance of the individual treatment of consumers. MC is also related to the increased global competition, shortening the life cycle of products as well as implementing new production and information technologies that enable companies to produce to customer specifications at low cost [4,5]. Facing the customer-driven market, the product design must cover a larger scope of the value chain and accentuate high-added value to the customers [6]. MC can be treated as a key instrument in building relations between producers and customers and gives manufacturers the opportunity to increase customer satisfaction, and hence customer retention in the long-term [7], which is followed by increasing customer loyalty [8].

According to the concept, mass personalization enables companies to achieve a competitive advantage through a product differentiation strategy while maintaining cost-effectiveness. Cavusoglu et al. [5] explained that if the customization cost is not low enough, companies should consider offering custom products instead of one single product. The authors emphasize that introducing flexibility requires a significant initial investment, known as the cost of operational flexibility. Furthermore, the mass production of tailor-made products cannot proceed directly without some loss of efficiency. Even in an increasingly individualized economy, many products are actually more semi-classic than completely custom-made. An example is a product in which the consumer can choose the species of wood, and the finish to be used in a particular design. Manufacturers of ready-to-assemble furniture are experiencing increasing demand from customers who expect products to be tailored to their specific needs such as a system for designing personalized ergonomic furniture (chairs, beds, tables, kitchen interiors, etc.) using anthropometric dimensions or other specific needs [9–12]. Based on the empirical test, Blecker and Abdelkafi [13] underlined that proliferation had a significant impact on cost due to the complexity of production, which affects the level of overhead costs. Of great importance, in addition to the variety of products, is the construction of the material being processed such as solid wood or wood-based panels, since the processing of wood-based materials is specific, which can sometimes limit the scope of customization [14,15]. As product variety increases, the planning complexity increases with more on-floor alternative routes, more work-in-progress inventory, assembly line balancing problems, increasing variability, etc. [13]. In the door industry, constraints on carrying out mass customization also arise from technical standards specifying requirements, for example, in terms of strength, intrusion resistance, soundproofing or other factors, with a particular focus on doors for public buildings [4].

Innovations in manufacturing processes should affect the cost reduction for the customer, reducing waste without additional resource requirements. Referring to the idea of a sustainable, circular economy, regeneration and the reuse of waste products and production residues can significantly reduce energy consumption and waste [16–18]. In modern production lines, achieving green production by saving energy and reducing emissions is possible by combining the automation of most production operations and the source and type of raw materials used in products such as using equally valuable forest biomass and its own production waste [16,19,20]. The answer to these requirements is lean manufacturing (LM) [21,22]. For any manufacturing company, machine breakdowns and downtime are a source of unavoidable costs. It is their reliability that affects the productivity of the company and directly affects the company's bottom line. More and more companies are recognizing the need to control the efficiency of machinery utilization, which allows them to identify waste in the technological processes implemented and the existing production reserves [23]. The target state that all enterprises should strive for is 100% utilization of the machinery park in their possession, and at the same time, no shortage of production is realized with an efficiency corresponding to the nominal efficiency of the technological equipment and machinery in their possession [24]. Based on these assumptions, the Porta KMI Poland door manufacturing plant is betting on innovation in the development and modernization of the production hall by creating a new customized TechnoPORTA technological line [25]. This is a fully automated intelligent module line that ensures the highest quality and production efficiency. Designed for mass customization while maintaining the required minimum production batch size (one door leaf), the line meets the highest technical requirements [4]. The line was developed using technologically advanced machinery and equipment, enabling the greatest possible automation of work at each stage of processing including the positioning and feeding of material and machine changeover [26,27].

The need for improvement in companies has existed for a long time, but in today's rapidly changing market, characterized by high dynamics and the need to flexibly adapt production to the needs, it is becoming essential to implement [28–31]. There are many tools and indicators for this to analyze the possibility of improving production processes in a manufacturing plant including TQM (total quality management) [32–34], lean management [35,36], Six Sigma [37,38], and quality management system (QMS) in accordance with the ISO standards [39,40]. An essential support for process improvement is the provision of adequate human resources to effectively manage and improve processes [41,42]. Achieving a high level of product quality requires continuous monitoring, analysis, and improvement of the process. Continuous monitoring of the production process involves recording and collecting accurate data on its progress. This type of activity is an important part of a company's management strategy [43]. For this purpose, it is possible to use manual methods (i.e., filling out the appropriate index cards or forms), but especially electronic methods (i.e., using MES—manufacturing execution systems). However, in order to analyze the process, it can be difficult to coordinate the use of measurements with different specifications and simultaneously from multiple workstations. It is much more practical to use numerical indicators of a synthetic nature that combine data from different sources. For this purpose, so-called key performance indicators (KPIs) are used in manufacturing systems. KPIs are defined as a set of measures (metrics) used to facilitate the evaluation of the performance of a production system from the perspective of productivity, quality, and maintenance [44,45].

In light of the need to evaluate the use of modules in the TechnoPORTA line in order to determine the fulfilment of the productivity design assumptions under the conditions specified by the MC concept, it was decided to conduct research to obtain production data from the IT system controlling the operation of the line. The aim of this research was to experimentally determine the performance of the edge banding module in the TechnoPORTA line on particular working days and to determine the possible influence of organizational and technological factors characterizing the line's operation, which can be defined and determined by analyzing the temporal technological data obtained from the IT system controlling the line's operation.

2. Materials and Methods

The edge banding module, which is crucial to the performance of the entire TechnoPORTA line was selected for testing. During the production process, each door leaf passes through this module at least three times (up to six times) to process three edges (two sides and one top). Therefore in order to produce one door leaf, the module must perform from three to six working cycles. In this process, subsequent machining units perform the following actions:

1. Feeding door leaves for production (feeding portal);
2. Reference edge milling (reference milling machine);
3. Edge processing (the edge with a rebate processing machine);
4. Stacking and transfer to the further processing of door leaves (two-way stacking portal).

The arrangement diagram and the most important machine units are shown in Figures 1 and 2.

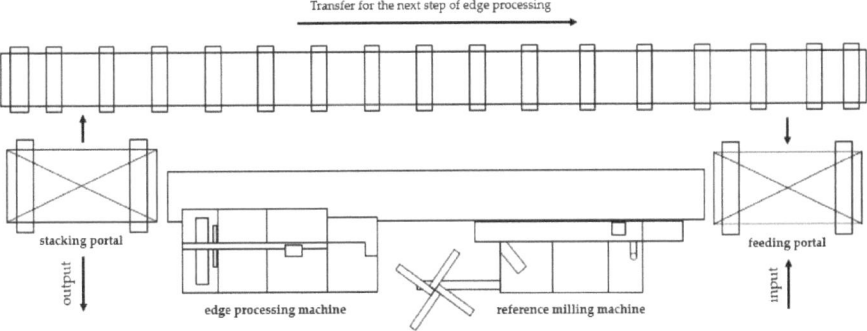

Figure 1. The arrangement diagram and the most important machine units of the edge banding module.

Figure 2. The units of the edge banding module: (**A**) feeding portal, (**B**) reference milling machine, (**C**) edge processing machine, (**D**) two-way stacking portal.

The research process consisted of two stages. In the first stage, machining data were collected from 14 September to 5 December 2020. Then, the data obtained from the IT system controlling the operation of the line were in the form of *.csv files. These data were analyzed in MS Excel software to determine the efficiency of the edge banding module.

A sample layout of machine data in a *.csv file is shown in Figure 3.

First, the data were sorted based on the "CYCLE" column (Figure 4). When the value in the column changed from "1" to "3", for example, it means that an entire pallet of one type of door leaf has gone through the machine tools to machine one edge, and then the machines are retooled to another type of door leaves ready on the next pallet.

Using the "IF(C1 <> C2; 1; 1; 0)" function, the data from the "CYCLE" column can be used to determine the moment when the machine retooling occurs. Passing through the first door leaf after retooling, it is marked with the number 1 in the "L" column. Then, in the "M" column, based on the "L" column, each series of processed door leaves is sequentially numbered thanks to the formula "=L2 + M1".

	A	B	C	D	E	F	G	H	I	J	K
1	ID	Production_Barcode_ID	CYCLE	maxCYCLE	Status	Date	User	Date	Time	Tact	Retooling time
2	200394	123065054	2	3	10	2020-09-14 08:40:10.05	JB	14.09.2020	8:40:10	0:01:25	#NAME?
3	200395	123065055	2	3	10	2020-09-14 08:41:35.75	JB	14.09.2020	8:41:35	0:01:19	#NAME?
4	200396	123065056	2	3	10	2020-09-14 08:42:54.253	JB	14.09.2020	8:42:54	0:00:11	#NAME?
5	200397	123065057	2	3	10	2020-09-14 08:43:05.0	JB	14.09.2020	8:43:05	0:01:30	#NAME?
6	200398	123065059	2	3	10	2020-09-14 08:44:35.897	JB	14.09.2020	8:44:35	0:00:11	#NAME?
7	200399	123065060	2	3	10	2020-09-14 08:44:46.05	JB	14.09.2020	8:44:46	0:00:10	#NAME?
8	200400	123065061	2	3	10	2020-09-14 08:44:56.553	JB	14.09.2020	8:44:56	0:20:32	0:20:32
9	200401	123065062	2	3	10	2020-09-14 09:05:28.21	JB	14.09.2020	9:05:28	0:07:51	0:07:51
10	200402	123065063	2	3	10	2020-09-14 09:13:19.41	JB	14.09.2020	9:13:19	0:01:12	#NAME?
11	200403	123065036	2	3	10	2020-09-14 09:14:31.493	JB	14.09.2020	9:14:31	0:01:36	#NAME?
12	200404	123065037	2	3	10	2020-09-14 09:16:07.557	JB	14.09.2020	9:16:07	0:00:10	#NAME?
13	200405	123065038	2	3	10	2020-09-14 09:16:17.75	JB	14.09.2020	9:16:17	0:00:11	#NAME?
14	200406	123065039	2	3	10	2020-09-14 09:16:28.207	JB	14.09.2020	9:16:28	0:02:34	#NAME?
15	200407	123065040	2	3	10	2020-09-14 09:19:02.607	JB	14.09.2020	9:19:02	0:00:36	#NAME?
16	200408	123065048	2	3	10	2020-09-14 09:19:38.497	JB	14.09.2020	9:19:38	0:00:09	#NAME?
17	200409	123065049	2	3	10	2020-09-14 09:19:47.943	JB	14.09.2020	9:19:47	0:01:17	#NAME?
18	200410	123065050	2	3	10	2020-09-14 09:21:04.95	JB	14.09.2020	9:21:04	0:00:11	#NAME?
19	200411	123065051	2	3	10	2020-09-14 09:21:15.053	JB	14.09.2020	9:21:15	0:00:10	#NAME?
20	200412	123065052	2	3	10	2020-09-14 09:21:25.35	JB	14.09.2020	9:21:25	0:00:16	#NAME?
21	200413	123065023	2	3	10	2020-09-14 09:21:41.95	JB	14.09.2020	9:21:41	0:00:11	#NAME?
22	200414	123065024	2	3	10	2020-09-14 09:21:52.25	JB	14.09.2020	9:21:52	0:00:10	#NAME?
23	200415	123065025	2	3	10	2020-09-14 09:22:02.2	JB	14.09.2020	9:22:02	0:00:10	#NAME?
24	200416	123065026	2	3	10	2020-09-14 09:22:12.957	JB	14.09.2020	9:22:12	0:02:08	#NAME?
25	200417	123065027	2	3	10	2020-09-14 09:24:20.403	JB	14.09.2020	9:24:20	0:00:12	#NAME?
26	200418	123065041	2	3	10	2020-09-14 09:24:32.603	JB	14.09.2020	9:24:32	0:00:10	#NAME?
27	200419	123065042	2	3	10	2020-09-14 09:24:42.557	JB	14.09.2020	9:24:42	0:00:11	#NAME?
28	200420	123065043	2	3	10	2020-09-14 09:24:53.153	JB	14.09.2020	9:24:53	0:00:10	#NAME?
29	200421	123065044	2	3	10	2020-09-14 09:25:03.16	JB	14.09.2020	9:25:03	0:00:10	#NAME?
30	200422	123065045	2	3	10	2020-09-14 09:25:13.5	JB	14.09.2020	9:25:13	0:00:49	#NAME?
31	200423	123065028	2	3	10	2020-09-14 09:26:02.85	JB	14.09.2020	9:26:02	0:00:11	#NAME?
32	200424	123065029	2	3	10	2020-09-14 09:26:13.35	JB	14.09.2020	9:26:13	0:00:10	#NAME?
33	200425	123065030	2	3	10	2020-09-14 09:26:23.1	JB	14.09.2020	9:26:23	0:00:10	#NAME?
34	200426	123065031	2	3	10	2020-09-14 09:26:33.497	JB	14.09.2020	9:26:33	0:00:10	#NAME?
35	200427	123065032	2	3	10	2020-09-14 09:26:43.843	JB	14.09.2020	9:26:43	0:00:18	#NAME?
36	200428	123065082	2	3	10	2020-09-14 09:27:01.197	JB	14.09.2020	9:27:01	0:00:10	#NAME?
37	200429	123065083	2	3	10	2020-09-14 09:27:11.397	JB	14.09.2020	9:27:11	0:00:10	#NAME?
38	200430	123065084	2	3	10	2020-09-14 09:27:21.2	JB	14.09.2020	9:27:21	0:00:10	#NAME?
39	200431	123065085	2	3	10	2020-09-14 09:27:31.593	JB	14.09.2020	9:27:31	0:00:10	#NAME?
40	200432	123065086	2	3	10	2020-09-14 09:27:41.543	JB	14.09.2020	9:27:41	0:00:10	#NAME?

Figure 3. An example of the machine report.

	A	B	C	D	E	F	G	H	I	J	K	L	M	N	O	P	Q	R	S
1	D	Production_Barcode_ID	CYCLE	maxCYCLE	Status	Date	User	Date	Time	Tact	Retooling time					Retooling y/n	Key	Max. Tact	Count
2	26827	123093324	2	6	10	2020-10-19 13:24:54.54	JB	19.10.2020	13:24:54	0:09:09	#NÁZEV?	1	1	1	1	1	1y	0:09:09	1
3	26828	123357016	1	3	10	2020-10-19 13:34:03.303	JB	19.10.2020	13:34:03	0:00:12	#NÁZEV?	1	2	5	5	0	2y	0:00:12	0
4	26829	123357015	1	3	10	2020-10-19 13:34:15.243	JB	19.10.2020	13:34:15	0:00:09	#NÁZEV?	0	2	3	5	0			0
5	26830	123393375	1	3	10	2020-10-19 13:34:24.69	JB	19.10.2020	13:34:24	0:00:12	#NÁZEV?	0	2	1	5	0			0
6	26831	123393376	1	3	10	2020-10-19 13:34:36.393	JB	19.10.2020	13:34:36	0:00:12	#NÁZEV?	0	2	5	5	0	2y	0:00:12	0
7	26832	123393385	1	3	10	2020-10-19 13:34:48.337	JB	19.10.2020	13:34:48	0:01:03	#NÁZEV?	0	2	1	5	0			0
8	26833	123225209	1	3	10	2020-10-19 13:35:51.39	JB	19.10.2020	13:35:51	0:00:12	#NÁZEV?	0	2	5	5	0	2y	0:00:12	0
9	26834	123225208	1	3	10	2020-10-19 13:36:03.293	JB	19.10.2020	13:36:03	0:00:09	#NÁZEV?	0	2	3	5	0			0
10	26835	123225203	1	3	10	2020-10-19 13:36:12.287	JB	19.10.2020	13:36:12	0:00:12	#NÁZEV?	0	2	5	5	0	2y	0:00:12	0
11	26836	123225202	1	3	10	2020-10-19 13:36:24.297	JB	19.10.2020	13:36:24	0:00:09	#NÁZEV?	0	2	3	5	0			0
12	26837	123225201	1	3	10	2020-10-19 13:36:33.287	JB	19.10.2020	13:36:33	0:00:12	#NÁZEV?	0	2	5	5	0	2y	0:00:12	0
13	26838	123225200	1	3	10	2020-10-19 13:36:45.19	JB	19.10.2020	13:36:45	0:00:09	#NÁZEV?	0	2	1	5	0			0
14	26839	123225199	1	3	10	2020-10-19 13:36:54.197	JB	19.10.2020	13:36:54	0:08:30	#NÁZEV?	0	2			1	2y	0:08:30	1
15	26840	123256303	3	6	10	2020-10-19 13:45:24.8	JB	19.10.2020	13:45:24	0:00:10	#NÁZEV?	1	3	1	1	0	3y	0:00:10	0
16	26841	123256304	3	6	10	2020-10-19 13:45:34.94	JB	19.10.2020	13:45:34	0:00:11	#NÁZEV?	0	3	1	1	0	3y	0:00:11	0
17	26842	123256305	3	6	10	2020-10-19 13:45:45.09	JB	19.10.2020	13:45:45	0:00:10	#NÁZEV?	0	3	1	1	0	3y	0:00:10	0
18	26843	123256306	3	6	10	2020-10-19 13:45:55.25	JB	19.10.2020	13:45:55	0:01:57	#NÁZEV?	0	3	1	1	0	3y	0:01:57	0
19	26844	123093324	3	6	10	2020-10-19 13:47:52.297	JB	19.10.2020	13:47:52	0:03:30	#NÁZEV?	0	3			1	3y	0:03:30	2
20	26845	123225199	2	3	10	2020-10-19 13:51:22.397	JB	19.10.2020	13:51:22	0:00:34	#NÁZEV?	1	4	1	2	0			0
21	26846	123225200	2	3	10	2020-10-19 13:51:56.54	JB	19.10.2020	13:51:56	0:00:10	#NÁZEV?	0	4	2	2	0	4y	0:00:10	0
22	26847	123225201	2	3	10	2020-10-19 13:52:06.537	JB	19.10.2020	13:52:06	0:00:10	#NÁZEV?	0	4	1	2	0			0
23	26848	123225202	2	3	10	2020-10-19 13:52:16.543	JB	19.10.2020	13:52:16	0:00:10	#NÁZEV?	0	4	2	2	0	4y	0:00:10	0
24	26849	123225203	2	3	10	2020-10-19 13:52:26.943	JB	19.10.2020	13:52:26	0:00:10	#NÁZEV?	0	4	2	2	0	4y	0:00:10	0
25	26850	123225208	2	3	10	2020-10-19 13:52:36.893	JB	19.10.2020	13:52:36	0:00:11	#NÁZEV?	0	4	1	2	0			0
26	26851	123225209	2	3	10	2020-10-19 13:52:47.09	JB	19.10.2020	13:52:47	0:03:39	#NÁZEV?	0	4	1	2	0			0
27	26852	123393385	2	3	10	2020-10-19 13:56:26.55	JB	19.10.2020	13:56:26	0:00:10	#NÁZEV?	0	4	2	2	0	4y	0:00:10	0
28	26853	123393376	2	3	10	2020-10-19 13:56:36.69	JB	19.10.2020	13:56:36	0:04:03	#NÁZEV?	0	4	1	2	0			0
29	26854	123393375	2	3	10	2020-10-19 14:00:39.193	JB	19.10.2020	14:00:39	0:00:35	#NÁZEV?	0	4	1	2	0			0
30	26855	123357015	2	3	10	2020-10-19 14:01:14.697	JB	19.10.2020	14:01:14	0:03:10	#NÁZEV?	0	4	1	2	0			0
31	26856	123357016	2	3	10	2020-10-19 14:04:24.04	JB	19.10.2020	14:04:24	0:03:18	#NÁZEV?	0	4			1	4y	0:03:18	1

Figure 4. An example of the machine data sorted by the column "CYCLE".

Based on the obtained data in column F (Date), the start and end time of the work shift could be determined. Based on these times, the production time for each day of door leaves production was calculated. From column B (Production Barcode ID), it was possible to count the number of working cycles performed on the door leaves produced each production day. These two parameters are very important for the main parameter characterizing production (i.e., the number of cycles performed per minute).

Column J shows the times of a working cycle for the production of door leaves. From these data, it was possible to calculate the average production time per working cycle using the function "=AVERAGE(Jx;Jy)". Using the function "=MODESNGL(Jx;Jy)" it was possible to find the most frequently occurring working cycle time, and using the function "=COUNTIF(Jx:Jy;most frequent time)", we could count how many times this time was reached during production.

Column K shows the retooling time. Using the filter, it was possible to extract and display only those rows containing times when retooling occurred. From the data displayed in this way, it was possible to calculate the average time for retooling for individual production days "=AVERAGE(Kx;Ky)" and the number of retooling occurred "=COUNT(Kx;Ky)".

All of these parameters (mean time of production per one working cycle, mean time of retooling, number of retooling, number of door leaves leaving in a series, most frequent time of series) were summarized in a sheet (Figure 5) and then processed in the form of graphs.

Date	Start of production	End of production	Time of production	Convert to minutes	Number of doors on the pallet	Number of door leaves	Number of cycles performed per min	Mean time of the cycle	Convert to secondes	Retooling time	Convert to minutes	Number of retoolings	Number of most frequent time	Convert to minutes	Most frequent time in %
14.09.2020	8:40:10	13:47:31	5:07:21	307.35000	20.26667	265	0.86221	0:01:10	70.00000	0:17:08	17.12576	10	4	0.66667	0.21691
15.09.2020	6:38:33	18:43:53	12:05:20	725.33333	40.66667	1017	1.40211	0:00:43	42.79253	0:15:06	15.10000	27	13	2.16667	0.29871
16.09.2020	8:23:10	18:39:04	10:15:54	615.90000	32.40909	714	1.15928	0:00:52	51.75630	0:15:48	15.80079	20	13	2.16667	0.35179
17.09.2020	8:18:55	18:46:52	10:27:57	627.95000	23.29730	863	1.37431	0:00:44	43.65817	0:15:19	15.31667	21	15	2.50000	0.39812
18.09.2020	6:43:25	18:34:57	11:51:32	711.53333	22.67241	1286	1.80736	0:00:33	33.19751	0:10:52	10.86746	21	27	4.50000	0.63244
19.09.2020	7:15:11	13:39:31	6:24:20	384.33333	40.50000	1003	2.60971	0:00:23	22.99103	0:11:47	11.78704	9	3	0.50000	0.13010
21.09.2020	6:48:19	13:47:09	6:58:50	418.83333	14.16393	865	2.06526	0:00:29	29.05202	0:11:08	11.13030	11	30	5.00000	1.19379
22.09.2020	6:44:31	12:32:00	5:47:29	347.48333	19.12500	919	2.64473	0:00:23	22.68662	0:08:39	8.65278	6	25	4.16667	1.19910
23.09.2020	7:04:53	13:40:58	6:36:05	396.08333	35.84000	876	2.21166	0:00:27	27.12900	0:09:30	9.49861	12	12	2.00000	0.50494
24.09.2020	6:43:16	13:52:48	7:09:32	429.53333	19.66667	1025	2.38631	0:00:25	25.14341	0:15:14	15.23095	7	33	5.50000	1.28046
25.09.2020	8:10:27	20:58:18	12:47:51	767.85000	20.51282	801	1.04317	0:00:58	57.51685	0:11:43	11.71838	39	12	2.00000	0.26047
28.09.2020	6:49:54	19:19:36	12:29:42	749.70000	22.68182	1498	1.99813	0:00:30	30.02804	0:13:35	13.58704	18	38	6.33333	0.84478
29.09.2020	6:45:51	18:00:43	11:14:52	674.86667	31.41379	912	1.35138	0:00:44	44.39912	0:14:27	14.45287	29	8	1.33333	0.19757
30.09.2020	6:47:11	21:04:13	14:17:02	857.03333	22.70588	1159	1.35234	0:00:44	44.36756	0:12:41	12.68542	32	29	4.83333	0.56396
01.10.2020	7:08:45	13:51:09	6:42:24	402.40000	20.00000	541	1.34443	0:00:45	44.62847	0:16:15	16.24394	11	9	1.50000	0.37276
02.10.2020	6:34:01	13:40:44	7:06:43	426.71667	19.61702	923	2.16303	0:00:28	27.73889	0:15:07	15.11042	8	26	4.33333	1.01551
05.10.2020	6:38:47	13:51:13	7:12:26	432.43333	28.62500	628	1.45225	0:00:41	41.31529	0:22:36	22.59500	10	6	1.00000	0.23125
06.10.2020	6:49:37	15:48:09	8:58:32	538.53333	18.87500	949	1.76219	0:00:34	34.04847	0:19:14	19.23889	12	20	3.33333	0.61897
07.10.2020	7:22:10	15:45:41	8:23:31	503.51667	20.36585	854	1.69607	0:00:35	35.37588	0:10:55	10.91905	21	23	3.83333	0.76131
08.10.2020	6:30:23	21:32:09	15:01:46	901.76667	17.20000	1807	2.00384	0:00:30	29.94245	0:10:39	10.65402	29	54	9.00000	0.99804
09.10.2020	6:33:00	21:19:55	14:46:55	886.91667	15.73239	1118	1.26055	0:00:48	47.59839	0:22:05	22.08788	22	31	5.16667	0.58254
12.10.2020	6:49:01	13:48:10	6:59:09	419.15000	46.52174	1071	2.55517	0:00:23	23.48179	0:13:06	13.10185	9	14	2.33333	0.55668
13.10.2020	6:51:51	13:46:14	6:54:23	414.38333	18.80000	754	1.81957	0:00:33	32.97480	0:10:04	10.06458	16	8	1.33333	0.32176
14.10.2020	6:56:47	16:19:51	9:23:04	563.06667	19.80769	1031	1.83104	0:00:33	32.76819	0:12:06	12.10526	19	15	2.50000	0.44400
15.10.2020	6:49:41	21:10:39	14:20:58	860.96667	23.87143	1672	1.94200	0:00:31	30.89593	0:10:55	10.91905	21	14	2.33333	0.27101
16.10.2020	7:03:12	17:13:48	10:10:36	610.60000	26.14815	1413	2.31412	0:00:26	25.92781	0:09:02	9.03148	18	22	3.66667	0.60050
19.10.2020	6:58:32	21:26:37	14:28:05	868.0833	28.17647059	1438	1.656522991	0:00:36	36.22044506	0:13:05	13.08988	28	24	4	0.46078525

Figure 5. An example of the statistical parameters determined based on the machine data.

At the end of the analysis, another sheet was created in the form of a contingency table, in which the calculated statistical parameters were averaged for each production day in the analyzed period (Figure 6).

	Values			
Date	Average lowest time	Average highest time	Sum of number of door on pallet	Sum of time without retooling
September 14, 2020	0:00:58	0:12:52	304	0:13:28
September 15, 2020	0:00:17	0:11:50	976	0:38:01
September 16, 2020	0:00:09	0:14:46	713	0:19:54
September 17, 2020	0:00:20	0:08:22	862	0:20:52
September 18, 2020	0:00:20	0:05:22	1315	0:42:02
September 19, 2020	0:00:09	0:04:33	972	0:09:04
September 21, 2020	0:00:22	0:02:39	918	0:13:38
September 22, 2020	0:00:45	0:05:09	896	0:07:59
September 23, 2020	0:00:16	0:03:34	1003	0:19:50
September 24, 2020	0:00:41	0:08:30	800	0:36:29
September 25, 2020	0:00:31	0:03:01	864	0:25:33
September 28, 2020	0:00:37	0:05:19	1497	0:36:22
September 29, 2020	0:00:15	0:12:33	911	0:37:54
September 30, 2020	0:00:27	0:08:21	1158	1:11:23
SUM	0:00:26	0:06:36	13189	6:32:29

Figure 6. An example of the monthly averaged data.

The results were summarized within the months of the study period (broken down into data from September, October, November, and December).

3. Results

The number of working cycles per shift in the function of the mean time of the cycle shows the strict theoretical inversely proportional relationship (Figure 7). However, this was not the subject of this analysis. It is important to show that under favorable conditions, the edge banding module of the TechnoPORTA line could achieve an efficiency level of over 2.5 cycles per minute. However, it is often lower, sometimes only 0.5 cycles per minute. This means that there are factors that limit this performance. The scientific goal should be to identify these factors and the business goal to eliminate or reduce their impact.

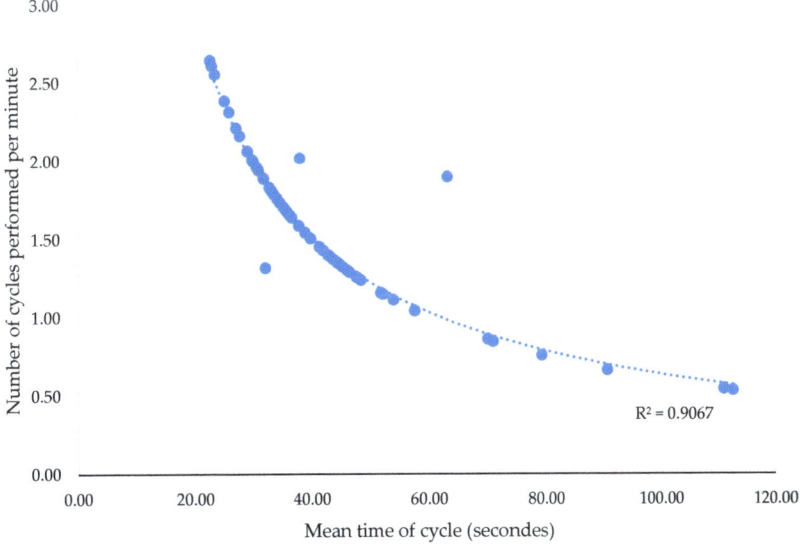

Figure 7. Dependence of the number of cycle doors per minute on the mean time of the cycle.

Figures 8–11 show that there is a low level of influence of internal factors, which can be determined based on the production data on the efficiency of this module. Figure 8 shows that the average mean time of retooling is between 10 and 20 min. The number of cycles performed per minute is not dependent on the average mean time of retooling. The number of retoolings from 5 to more than 40 did not change the number of cycles performed per minute, falling within the range of 0.5 to 2.5 per minute (Figure 9). Similarly, the number of doors per pallet, ranging from 10 to even 60, did not affect the efficiency (Figure 10). The most frequent processing times per day fell within a wide range from 3 to 54 cycles. Their number is irrelevant to the performance calculated in cycles per minute (Figure 11). Although the overall efficiency range was between 0.5 and 2.5 cycles per minute, on the vast majority of days, a narrower range of 1 to 2 cycles per minute was achieved, regardless of the values of the internal factors measured.

The mean time of retooling, number of retooling, number of doors on the pallet, and number of the most frequent time in one working shift did not influence the efficiency of the customized module of the edge banding of door leaves. It can be concluded that the machine data collected from 14 September to 5 December 2020 did not allow us to identify factors influencing the daily variability of efficiency. Either the IT system was unable to catch them in this time, or they were outside the area controlled by the system.

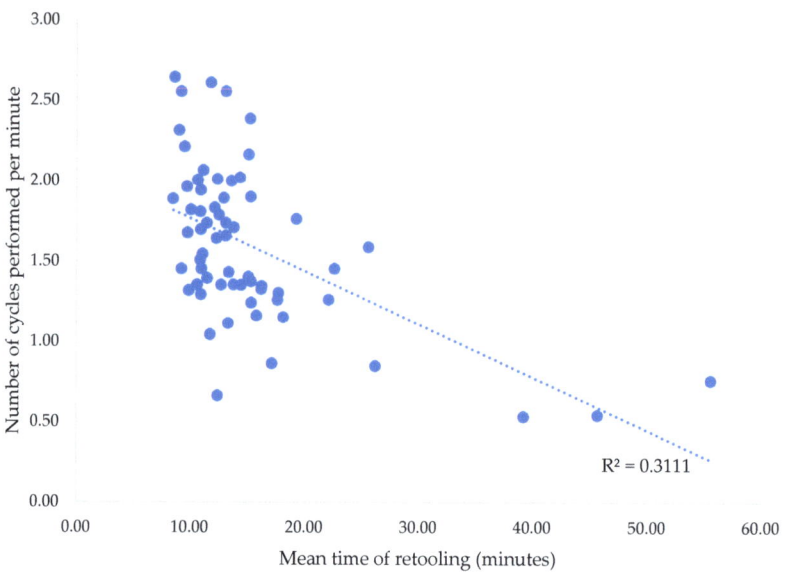

Figure 8. Dependence of the number of cycles performed per minute on the mean time of retooling.

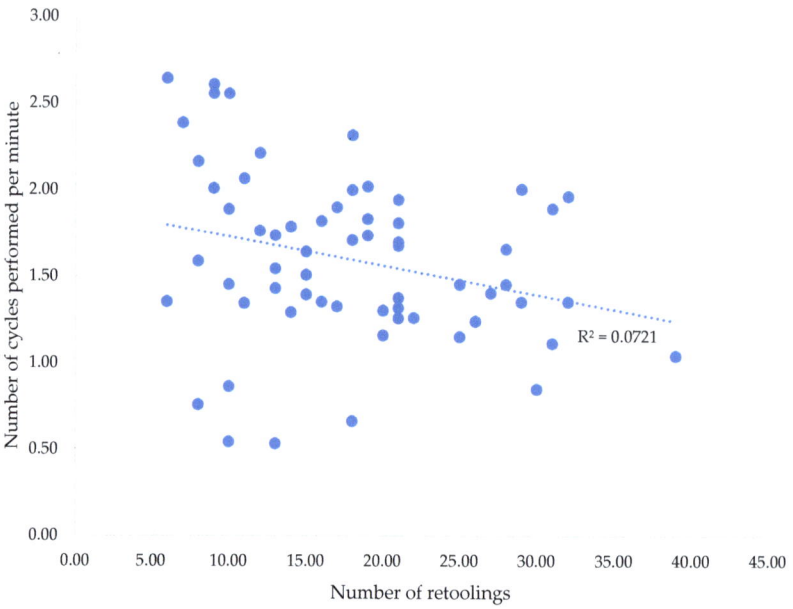

Figure 9. Dependence of the number of cycles performed per minute on the number of retoolings.

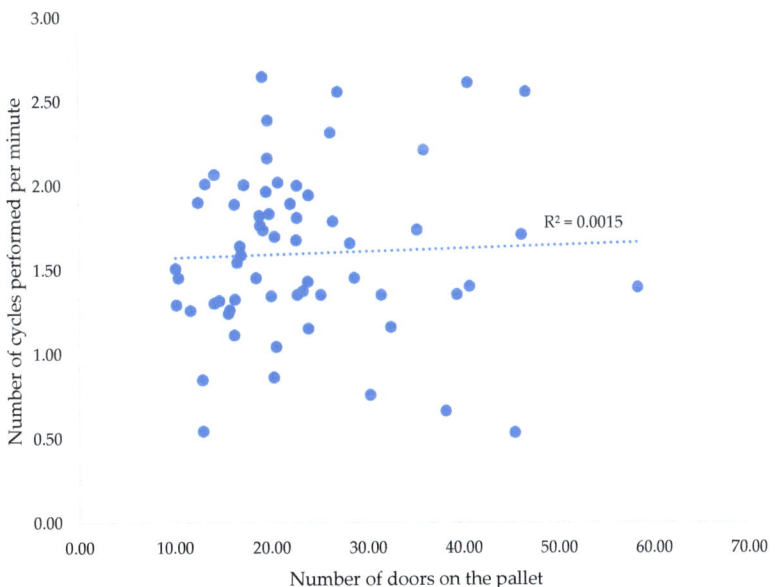

Figure 10. Dependence of the number of cycles performed per minute on the number of doors on the pallet.

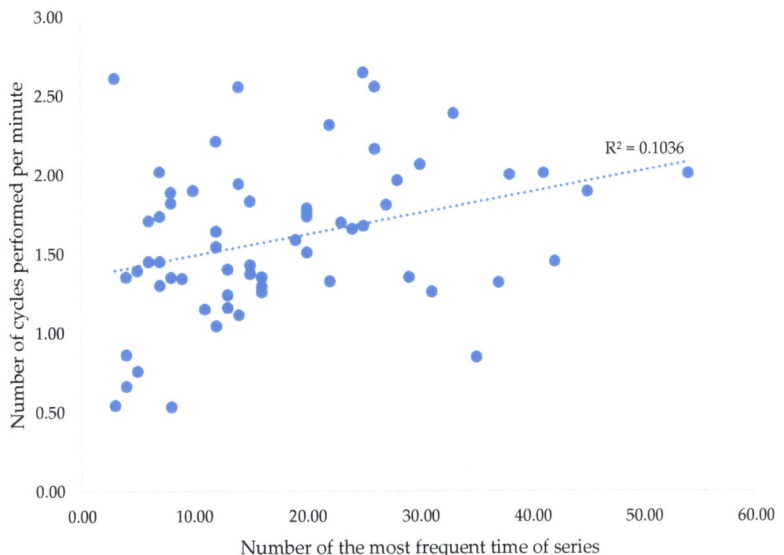

Figure 11. Dependence of the number of cycles performed per minute and the number of the most frequent time of series.

4. Discussion

Although MC is a trend that nowadays is developing dynamically, companies can perceive this phenomenon as a significant challenge. It happens because it requires both the products and the production processes to be customized [46].

There are two basic groups of limitations of mass customization connected with external (customer, market) and internal (production, logistic) elements, looking from the point of view of entities [47]. The first group of restrictions associated with external factors are related to the structure of the market and unfavorable demand conditions. It is fundamental for companies to determine whether there is a potential market demand big enough to respond to the customers' preferences and expectations. It is mostly connected with visible and measurable demand conditions that appear on the market. Customers demand variety when they differ sharply in their preferences for certain attributes of a product. Under such circumstances, customization may truly add value [5,47]. Production processes must be flexible to provide individual product features, bearing in mind the impact of the technological and material parameters on the final quality of, for example, the final processing surface during edge banding, milling, or other operations in individual technological line modules [48,49]. Milling is at the forefront of operations with which to give the final product the characteristics the customer wants. On the other hand, any slowdown or recession may reduce the rationality of the implementation of the customization process.

It is also worth noting that the changes on the market have a cyclical nature (the theory of business cycles). In particular, the concept of Kitchin's minor business cycle can be applied. described by Grasselli and Nguyen-Huu [50], taking a supply-side perspective, manufacturers decide on the level of production based on the expected level of demand and the desired level of inventory. On the demand side, investment is determined as a function of utilization and profitability and can be financed by debt, whereas consumption is independently determined as a function of income and wealth. In business activities, the time lags in information and decision adjustment affect the prices, output, demand, inventory, and employment in a periodic manner [50].

Mendelson and Parlaktürk [51] found that the implementation of MC depends on the company's competitive position on the market. They concluded that in a competitive environment, before introducing MC, a company should consider reducing costs or making efforts to improve the product quality. In the case of a monopoly, where price competition does not exist, the company can easily implement costless personalization. However, strong competition in the market makes companies accept worse financial results than if they offered standard products [5].

The internal adverse factors of companies are those connected with production, technologies, and logistics. Mass customization requires a highly flexible production technology [4]. Increasing flexibility takes place through the implemented innovations in the field of modular structures and cost-effective operations as well as the increasingly common use of information and digital technology to control production devices. Product–process–supply chain engineering is coexistent in MC. Manufacturers implement reconfigurable production systems (RMS) that help to operate in unpredictable and changeable markets through the machines' reconfigurability and flexibility [2]. Nevertheless, as described by Modrak and Soltysova [52], assembly line balancing causes difficult and important problems for manufacturers. Mass customization also requires a properly adapted logistics system for the customer [47].

Taking into account the above-mentioned external and internal factors, companies have to decide not only on whether to customize their products to the needs of the consumers, but also the level of this adaptation. As mentioned by Hou et al. [53], manufacturers can use optimization models to balance the costs of customization and the economies of the scale of mass production. In industry, only certain product features can be customized because only certain processing steps are sufficiently flexible.

The results of the study showed that neither the mean time of retooling, the number of retooling a working shift, the number of door leaves on the pallet, nor the number of the most frequent processing time in a door leaves series influenced the efficiency of the line. These technological and production organization parameters are the internal factors of the line, which can be calculated based on the machine reports generated and recorded in the IT production system. Therefore, it can be said that the edge banding module in the TechnoPORTA line is flexible in known and measurable conditions.

However, as a result of the tests, a significant variation in the line efficiency was found, defined by the average number of door leaves produced per minute, which was from just over 0.5 to over 2.5 pieces per minute. This proves the presence of external factors influencing the efficiency of the line, which could be connected with the customers' order structure or internal factors unidentified in the conducted tests. In this case, external factors may include the volatility of the structure characteristic of the customers' orders. On the other hand, the IT production control system was updated and reprogrammed to identify possible internal factors related to the operation of the TechnoPORTA line. The efficiency of the line was also further stabilized, but it is still necessary to carry out more extended working time observations, especially in terms of identifying the causes of downtime.

5. Conclusions

Based on the analysis of the obtained data and the discussion on the factors limiting mass customized production systems, it can be concluded that in the case of the edge banding module of the TechnoPORTA line, there was no influence of the mean time of retooling, the number of retooling, the number of doors on the pallet, and the number of the most frequent time in one working shift on the efficiency of this module. The data collected by the IT system controlling the line indicates that this module is flexible and its performance is not related to the control parameters. However, the large differences in efficiency on individual days indicate the existence of efficiency-limiting factors. Therefore, for the effective implementation of mass customization, thanks to which the efficiency of the line modules will be stable and their operation as flexible as possible, it is necessary to conduct planned technological tests to identify these factors and their significance. In order to identify them, the process engineering department reprogrammed the IT system to use production scheduling (ERP) data. Tests are in progress, which will show that the identification and control of the technological factors improve the performance of the customized line. The further direction of research will be the interpretation of the results obtained, and based on them, the determination of further work to improve the operation of the module and the replication of the work schedule to subsequent modules of the technological line.

Author Contributions: Conceptualization, Z.K., T.R., L.H. and E.T.; Methodology, Z.K., Ł.S., B.K., E.T., L.H. and J.P.; Software, Ł.S., B.K., L.H. and E.T.; Validation, Z.K., T.R. and M.P.; Formal analysis, Z.K., T.R. and L.H.; Investigation, Z.K. Ł.S., B.K., E.T. and T.R.; Resources, Z.K. Ł.S., B.K., J.B. and E.T.; Data curation, L.H., E.T. and J.P.; Writing—original draft preparation, T.R., J.B., L.H., E.T. and M.P.; Writing—review and editing, T.R. and M.P.; Visualization, L.H., Z.K., E.T. and M.P.; Supervision, T.R. and Z.K.; Project administration, Z.K. and T.R.; Funding acquisition, T.R. and Z.K. All authors have read and agreed to the published version of the manuscript.

Funding: This research received no external funding.

Institutional Review Board Statement: Not applicable.

Informed Consent Statement: Not applicable.

Data Availability Statement: Not applicable.

Conflicts of Interest: The authors declare no conflict of interest.

References

1. Davis, S.M. *Future Perfect*; Addison-Wesley Publishing: Reading, MA, USA, 1987.
2. Sabioni, R.C.; Daaboul, J.; le Duigou, J. Joint Optimization of Product Configuration and Process Planning in Reconfigurable Manufacturing Systems. *Int. J. Ind. Eng. Manag.* **2022**, *13*, 58–75. [CrossRef]
3. Piller, F.T. Mass Customization: Reflections on the State of the Concept. *Int. J. Flex. Manuf. Syst.* **2004**, *16*, 313–334. [CrossRef]
4. Pędzik, M.; Bednarz, J.; Kwidziński, Z.; Rogoziński, T.; Smardzewski, J. The Idea of Mass Customization in the Door Industry Using the Example of the Company Porta KMI Poland. *Sustainability* **2020**, *12*, 3788. [CrossRef]

5. Cavusoglu, H.; Cavusoglu, H.; Raghunathan, S. Selecting a Customization Strategy Under Competition: Mass Customization, Targeted Mass Customization, and Product Proliferation. *IEEE Trans. Eng. Manag.* **2007**, *54*, 12–28. [CrossRef]
6. Du, X.; Jiao, J.; Tseng, M.M. Understanding Customer Satisfaction in Product Customization. *Int. J. Adv. Manuf. Technol.* **2006**, *31*, 396–406. [CrossRef]
7. Bardakci, A.; Whitelock, J. Mass-customisation in Marketing: The Consumer Perspective. *J. Consum. Mark.* **2003**, *20*, 463–479. [CrossRef]
8. Wang, Y.; Wu, J.; Lin, L.; Shafiee, S. An Online Community-Based Dynamic Customisation Model: The Trade-off between Customer Satisfaction and Enterprise Profit. *Int. J. Prod. Res.* **2021**, *59*, 1–29. [CrossRef]
9. Bonenberg, A.; Branowski, B.; Kurczewski, P.; Lewandowska, A.; Sydor, M.; Torzyński, D.; Zabłocki, M. Designing for Human Use: Examples of Kitchen Interiors for Persons with Disability and Elderly People. *Hum. Factors Ergon. Manuf. Serv. Ind.* **2019**, *29*, 177–186. [CrossRef]
10. Gejdoš, M.; Hitka, M. The Impact of the Secular Trend of the Slovak Population on the Production of Wooden Beds and Seating Furniture. *Forests* **2022**, *13*, 1599. [CrossRef]
11. Bumgardner, M.S.; Nicholls, D.L. Sustainable Practices in Furniture Design: A Literature Study on Customization, Biomimicry, Competitiveness, and Product Communication. *Forests* **2020**, *11*, 1277. [CrossRef]
12. Langová, N.; Réh, R.; Igaz, R.; Krišťák, Ľ.; Hitka, M.; Joščák, P. Construction of Wood-Based Lamella for Increased Load on Seating Furniture. *Forests* **2019**, *10*, 525. [CrossRef]
13. Blecker, T.; Friedrich, G. *Mass Customization: Challenges and Solutions*; Springer: Boston, MA, USA, 2006; Volume 87, ISBN 978-0-387-32222-3.
14. Sydor, M.; Pinkowski, G.; Kučerka, M.; Kminiak, R.; Antov, P.; Rogoziński, T. Indentation Hardness and Elastic Recovery of Some Hardwood Species. *Appl. Sci.* **2022**, *12*, 5049. [CrossRef]
15. Sydor, M.; Rogoziński, T.; Stuper-Szablewska, K.; Starczewski, K. The Accuracy of Holes Drilled in the Side Surface of Plywood. *Bioresources* **2019**, *15*, 117–129. [CrossRef]
16. Yang, Y.; Yuan, G.; Zhuang, Q.; Tian, G. Multi-Objective Low-Carbon Disassembly Line Balancing for Agricultural Machinery Using MDFOA and Fuzzy AHP. *J. Clean. Prod.* **2019**, *233*, 1465–1474. [CrossRef]
17. Jiang, Z.; Wang, H.; Zhang, H.; Mendis, G.; Sutherland, J.W. Value Recovery Options Portfolio Optimization for Remanufacturing End of Life Product. *J. Clean. Prod.* **2019**, *210*, 419–431. [CrossRef]
18. Ghayebloo, S.; Tarokh, M.J.; Venkatadri, U.; Diallo, C. Developing a Bi-Objective Model of the Closed-Loop Supply Chain Network with Green Supplier Selection and Disassembly of Products: The Impact of Parts Reliability and Product Greenness on the Recovery Network. *J. Manuf. Syst.* **2015**, *36*, 76–86. [CrossRef]
19. Pędzik, M.; Tomczak, K.; Janiszewska-Latterini, D.; Tomczak, A.; Rogoziński, T. Management of Forest Residues as a Raw Material for the Production of Particleboards. *Forests* **2022**, *13*, 1933. [CrossRef]
20. Pędzik, M.; Kwidziński, Z.; Rogoziński, T. Particles from Residue Wood-Based Materials from Door Production as an Alternative Raw Material for Production of Particleboard. *Drv. Ind.* **2022**, *73*, 351–357. [CrossRef]
21. Bhamu, J.; Singh Sangwan, K. Lean Manufacturing: Literature Review and Research Issues. *Int. J. Oper. Prod. Manag.* **2014**, *34*, 876–940. [CrossRef]
22. Gupta, S.; Jain, S.K. A Literature Review of Lean Manufacturing. *Int. J. Manag. Sci. Eng. Manag.* **2013**, *8*, 241–249. [CrossRef]
23. Antosz, K.; Ciecińska, B. *Podstawy Zarządzania Parkiem Maszyn w Przedsiębiorstwie*; Oficyna Wydawnicza Politechniki Rzeszowskiej: Rzeszów, Poland, 2011; ISBN 978-83-7199-688-7.
24. Furman, J.; Orszag, P.R. Slower Productivity and Higher Inequality: Are They Related? *SSRN Electron. J.* **2018**. [CrossRef]
25. Kwidziński, Z.; Drewczyński, M.; Rogoziński, T.; Pędzik, M. Energy Efficiency in Mass Customized Production of Wooden Doors. In Proceedings of the Chip and Chipless Woodworking Processes, Zvolen, Slovakia, 15–17 September 2022; Volume 13.
26. Kwidziński, Z.; Bednarz, J.; Sankiewicz, Ł.; Pędzik, M.; Rogoziński, T. TechnoPORTA Intelligent, Customized Technological Line for the Automated Production of Technical Doors—Selected Technical and Economic Indicators. *Ann. WULS For. Wood Technol.* **2021**, *114*, 96–100. [CrossRef]
27. Kwidziński, Z.; Bednarz, J.; Pędzik, M.; Sankiewicz, Ł.; Szarowski, P.; Knitowski, B.; Rogoziński, T. Innovative Line for Door Production Technoporta—Technological and Economic Aspects of Application of Wood-Based Materials. *Appl. Sci.* **2021**, *11*, 4502. [CrossRef]
28. Liker, J.K. *The Toyota Way: 14 Management Principles from the World's Greatest Manufacturer*; McGraw-Hill: New York, NY, USA, 2004; ISBN 978-0-07-139231-0.
29. Womack, J.P.; Jones, D.T. *Lean Thinking*, 1st ed.; Lean Enterprise Institute; Taylor & Francis: Abingdon, UK, 1996; ISBN 978-0-684-81035-5.
30. Malms, M.; Ostasz, M.; Gilliot, M.; Bernier-Bruna, P.; Cargemel, L.; Suarez, E.; Cornelius, H.; Duranton, M.; Koren, B.; Rosse-Laurent, P.; et al. ETP4HPC's Strategic Research Agenda for High-Performance Computing in Europe 4. 2020. Available online: https://www.google.com.hk/url?sa=t&rct=j&q=&esrc=s&source=web&cd=&ved=2ahUKEwiUvLiBzeb7AhXWUGwGHQsJAyoQFnoECAwQAQ&url=https%3A%2F%2Fhal.inria.fr%2Fhal-03354396%2Ffile%2FETP4HPC_SRA4_2020_web_20200331.pdf&usg=AOvVaw1qFzhCZq0jSEhLZkZEAPGD (accessed on 15 November 2022). [CrossRef]
31. Skotnicka-Zasadzień, B. Analiza Efektywności Zastosowania Metody FMEA w Małym Przedsiębiorstwie Przemysłowym. *Syst. Support. Prod. Eng.* **2012**, *2*, 142–153.

32. Permana, A.; Purba, H.H.; Rizkiyah, N.D. A Systematic Literature Review of Total Quality Management (TQM) Implementation in the Organization. *Int. J. Prod. Manag. Eng.* **2021**, *9*, 25. [CrossRef]
33. Lasrado, F.; Nyadzayo, M. Improving Service Quality. *Int. J. Qual. Reliab. Manag.* **2020**, *37*, 393–410. [CrossRef]
34. Tasleem, M.; Khan, N.; Masood, S.A. Impact of TQM and Technology Management on Organizational Performance. *Mehran Univ. Res. J. Eng. Technol.* **2016**, *35*, 585–598. [CrossRef]
35. Garcia-Buendia, N.; Moyano-Fuentes, J.; Maqueira-Marín, J.M.; Cobo, M.J. 22 Years of Lean Supply Chain Management: A Science Mapping-Based Bibliometric Analysis. *Int. J. Prod. Res.* **2021**, *59*, 1901–1921. [CrossRef]
36. Uriona Maldonado, M.; Leusin, M.E.; de Albuquerque Bernardes, T.C.; Vaz, C.R. Similarities and Differences between Business Process Management and Lean Management. *Bus. Process Manag. J.* **2020**, *26*, 1807–1831. [CrossRef]
37. Simanova, Ľ.; Gejdoš, P. Implementation of the Six Sigma Methodology in Increasing the Capability of Processes in the Company of the Furniture Industry of the Slovak Republic. *Manag. Syst. Prod. Eng.* **2021**, *29*, 54–58. [CrossRef]
38. Raval, S.J.; Kant, R.; Shankar, R. Analyzing the Lean Six Sigma Enabled Organizational Performance to Enhance Operational Efficiency. *Benchmarking Int. J.* **2020**, *27*, 2401–2434. [CrossRef]
39. Sfreddo, L.S.; Vieira, G.B.B.; Vidor, G.; Santos, C.H.S. ISO 9001 Based Quality Management Systems and Organisational Performance: A Systematic Literature Review. *Total Qual. Manag. Bus. Excell.* **2021**, *32*, 389–409. [CrossRef]
40. Pacana, A.; Ulewicz, R. Analysis of Causes and Effects of Implementation of the Quality Management System Compliant with ISO 9001. *Pol. J. Manag. Stud.* **2020**, *21*, 283–296. [CrossRef]
41. Bessant, J.; Lamming, R.; Noke, H.; Phillips, W. Managing Innovation beyond the Steady State. *Technovation* **2005**, *25*, 1366–1376. [CrossRef]
42. Yen-Tsang, C.; Csillag, J.M.; Siegler, J. Theory of Reasoned Action for Continuous Improvement Capabilities: A Behavioral Approach. *Rev. Adm. Empres.* **2012**, *52*, 546–564. [CrossRef]
43. Corredera, A.; Macia, A.; Sanz, R.; Hernandez, J.L. An Automated Monitoring System for Surveillance and KPI Calculation. In Proceedings of the 2016 IEEE Workshop on Environmental, Energy, and Structural Monitoring Systems (EESMS), Bari, Italy, 13–14 June 2016; pp. 1–6.
44. Borsos, G.; Iacob, C.C.; Calefariu, G. The Use KPI's to Determine the Waste in Production Process. *IOP Conf. Ser. Mater. Sci. Eng.* **2016**, *161*, 012102. [CrossRef]
45. Tzeng, G.-H.; Cheng, H.-J.; Huang, T.D. Multi-Objective Optimal Planning for Designing Relief Delivery Systems. *Transp. Res. E Logist. Transp. Rev.* **2007**, *43*, 673–686. [CrossRef]
46. Yu, B.; Zhao, H.; Xue, D. A Multi-Population Co-Evolutionary Genetic Programming Approach for Optimal Mass Customisation Production. *Int. J. Prod. Res.* **2017**, *55*, 621–641. [CrossRef]
47. Zipkin, P. The Limits of Mass Customization. *Sloan Manag. Rev* **2001**, *42*, 81–87.
48. Korčok, M.; Koleda, P.; Barcík, Š.; Očkajová, A.; Kučerka, M. Effect of Technological and Material Parameters on Final Surface Quality of Machining When Milling Thermally Treated Spruce Wood. *Martin* **2019**, *14*, 10004–10013.
49. Kminiak, R.; Siklienka, M.; Igaz, R.; Krišťák, Ľ.; Gergeľ, T.; Němec, M.; Réh, R.; Očkajová, A.; Kučerka, M. Effect of Cutting Conditions on Quality of Milled Surface of Medium-Density Fibreboards. *Bioresources* **2019**, *15*, 746–766. [CrossRef]
50. Grasselli, M.R.; Nguyen-Huu, A. Inventory Growth Cycles with Debt-Financed Investment. *Struct. Chang. Econ. Dyn.* **2018**, *44*, 1–13. [CrossRef]
51. Mendelson, H.; Parlaktürk, A.K. Competitive Customization. *Manuf. Serv. Oper. Manag.* **2008**, *10*, 377–390. [CrossRef]
52. Modrak, V.; Soltysova, Z. Batch Size Optimization of Multi-Stage Flow Lines in Terms of Mass Customization. *Int. J. Simul. Model.* **2020**, *19*, 219–230. [CrossRef]
53. Hou, S.; Gao, J.; Wang, C. Design for Mass Customisation, Design for Manufacturing, and Design for Supply Chain: A Review of the Literature. *IET Collab. Intell. Manuf.* **2022**, *4*, 1–16. [CrossRef]

Article

Prediction of the Effect of CO_2 Laser Cutting Conditions on Spruce Wood Cut Characteristics Using an Artificial Neural Network

Ivan Ružiak *, Rastislav Igaz, Ivan Kubovský, Milada Gajtanska and Andrej Jankech

Faculty of Wood Sciences and Technology, Technical University in Zvolen, T. G. Masaryka 24, SK-960 01 Zvolen, Slovakia
* Correspondence: ruziak@tuzvo.sk

Abstract: In addition to traditional chip methods, performance lasers are often used in the field of wood processing. When cutting wood with CO_2 lasers, it is primarily the area of optimization of parameters that is important, which include mainly laser performance and cutting speed. They have a significant impact on the production efficiency and cut quality. The article deals with the use of an artificial neural network (ANN) to predict spruce wood cut characteristics using CO_2 lasers under several conditions. The mutual impact of the laser performance (P) and the number of annual circles (AR) for prediction of the characteristics of the cutting kerf and the heat affected zone (HAZ) were examined. For this purpose, the artificial neural network in Statistica 12 software was used. The predicted parameters can be used to qualitatively characterize the cutting kerf properties of the spruce wood cut by CO_2 lasers. All the predictions are in good agreement with the results from the available literary sources. The laser power P = 200 W provides a good cutting quality in terms of cutting kerf widths ratio defined as the ratio of cutting kerf width at the lower board to the cutting kerf width at upper board and, therefore, they are optimal for cutting spruce wood at $1.2 \cdot 10^{-2}$ m·s^{-1}.

Keywords: CO_2 laser; artificial neural networks; wood kerf; spruce wood; heat affected zone

Citation: Ružiak, I.; Igaz, R.; Kubovský, I.; Gajtanska, M.; Jankech, A. Prediction of the Effect of CO_2 Laser Cutting Conditions on Spruce Wood Cut Characteristics Using an Artificial Neural Network. *Appl. Sci.* 2022, *12*, 11355. https://doi.org/10.3390/app122211355

Academic Editor: Giuseppe Lazzara

Received: 21 October 2022
Accepted: 6 November 2022
Published: 9 November 2022

Publisher's Note: MDPI stays neutral with regard to jurisdictional claims in published maps and institutional affiliations.

Copyright: © 2022 by the authors. Licensee MDPI, Basel, Switzerland. This article is an open access article distributed under the terms and conditions of the Creative Commons Attribution (CC BY) license (https://creativecommons.org/licenses/by/4.0/).

1. Introduction

Wood cutting by CO_2 lasers is one of the basic methods of wood cutting to relatively cut any dimensions and shapes. Among their advantages in comparison to other cutting techniques such as CNC processing and water jet processing, belong quickness, non-contact, good wood cutting, efficiency and therefore also the good surface properties of the wood. Another advantage also lies in fact that by changing the parameters of the laser ray such as power, cutting speed, cutting angle, protective atmosphere and its pressure, it is possible to reach any sample dimensions with a good quality of surface which is crucial for the next step of processing the sample. Another advantage of laser cutting lies in the fact that lasers affect only limited areas by thermal stress. [1–6].

In the world of science there are many articles which deal with the effect of CO_2 laser parameters and their effect on wood cut parameters. Factors which influence final the cut of wood can be divided into three groups which are properties of the radiation beam, properties of the laser device and the characteristics of the cutting process such as laser power P, cutting speed v, number of annual rings AR, moisture, focal point position and many more. The effect of P, v, and AR on the cut characteristics of spruce wood was studied in [7].

Nukman et al. [8] studied the same cut parameters for Malaysian-based woods and plywood and also presented the dependence of material removal rate MRR vs. P, v from which it is good to see that the MRR increases with the laser power in the exponential stabilizing form in the compressed air or nitrogen atmosphere. Other authors have studied the effect of CO$_2$ laser parameters on the width of the heat affected zone in wood-based materials and wood composites. [9].

Th effect of laser parameters on the cut properties of wood-based composite materials such as MDF were studied in [10]. The authors studied MDF cutting by a CO$_2$ laser for P = 520–530 W in the pulse and continuous mode. They found out that narrow kerf widths can be achieved for the pulse mode.

Eltawahni et al. [11] defined methodology for evaluation of radiation beam efficiency and quality based on the values of parameter WKR defined as the ratio of cutting kerf width at the lower board WKL vs. the cutting kerf width at the upper board WKU (WKR = WKL/WKU) which is mainly affected by P, v and position of the focal point. They also find out that roughness increases with cutting speed and air pressure and decreases with P and focal point position. In other work Eltawahni et al. [12] studied the effect of laser parameters on the final cut of plywood materials.

Many authors have also studied the effect of focal point position on the final cut properties. For wood-based materials the results are listed in [13] and for wood composites they are listed in [14,15]. The effect of moisture content on the cutting of pine wood by CO$_2$ lasers was studied in [16]. The effect of assistance gas on the laser cutting of wood-based materials was studied in [17] and the effect of processing parameters and the parameters of gas on the cutting of micro thin wood was studied in [18].

Values of cut parameters are often processed by K—means clustering algorithm from the measured picture of the surface by several types of microscopies which is described in [19].

CO$_2$ lasers together with UV lasers, VIS and NIR lasers offer similar results in cutting. All these lasers offer better results than other techniques. Several authors use UV, VIS and NIR lasers for wood cutting. The next part deals with the effect of such laser's parameters on the machining of wood and wood composites.

Authors in the work [20] have studied the effect of wavelength and pulse width on the cut of Japanese larch, cedar and beech wood using UV laser. In the work [21], the authors have analyzed the effect of UV, VIS and NIR laser wavelengths on the machining performance of wood.

The effect of laser properties on the surface characteristics of different types of woods was studied in the works [22–25]. The authors of the work [22] studied the surface properties of beech wood after CO$_2$ laser engraving. In the previous work [23] the authors studied impact of radiation forms on beech wood color changes. The effect of CO$_2$ laser parameters on color changes in hardwoods and in limewood was published in the works [24,25].

Artificial neural networks have found very broad usage in material science such as wood science, polymers science, metals science but also optimization of technologies for materials processing. Many authors have used this very useful method for predicting wood materials properties. Authors of the work [26] have predicted thermal conductivity of wood using artificial neural networks. Surface roughness of wood in machining process was modelled in [27] and ANN was applied for minimalizing surface roughness and power consumption in abrasive machining of wood in the work [28]. ANN was also applied for prediction of optimum power consumption in wood machining in the work [29].

Neural networks have also been widely used for optimization of materials technologies. The authors in the work [30] modelled formaldehyde emissions during a particleboard manufacturing process and published the effect of the manufacturing technology on the modulus of rupture (MOR) and the modulus of elasticity (MOE) vs. pressing conditions in the work [31].

Other authors have dealt with prediction of adhesive bonding strength and bonding quality vs. pressing conditions using ANN or multiple linear regression models [32–34].

Artificial neural networks have also been used for prediction of color changes in wood which are largely affected by radiation from CO$_2$ or other types of lasers. The authors in the work [35] predicted the color changes of heat-treated wood during artificial weathering and during heat treatment by natural atmospheric conditions in the work [36].

The authors of the work [37] used ANN for determination of CNC processing parameters for the best wood surface quality. In the article [38], they determined the surface properties of MDF and optimized CNC processing parameters for this type of wood composite.

In comparison to MLP and RBF networks, artificial neural networks offer other possibilities for prediction of manufacturing industry demands using a deep learning approach. This method was described in detail in [39].

The goal of the article was to the predict values of cutting parameters at non-measured values of laser power P. These results can be successfully used for complex characterization of P, the AR effect on cut parameters for all applicable values of laser power for Picea abies. The results of ANN can be used for prediction of cut parameters at any laser power values between 100 and 500 W for all possible values of AR.

2. Materials and Methods

The experiments were carried on spruce wood (*Picea abies* L.). Experimental equipment LCS 400 (VEB Feinmechanische Werke, Halle, Germany) was used for cutting. This system consists of a CO_2 laser (wavelength 10.6 µm and maximum power output 400 W), a positioning table system (laser head positioning in the plane formed by the x and y axes) and a special PC control system. The sample was placed in lens focus. The focused laser beam stroked perpendicularly on the surface of the sample and the laser head carriage moved along the width (axis x) at a certain scanning speed until the wood sample was cut off. Cutting kerf was obtained from the wood specimens with dimensions T × R × L (8 × 100 × 1000) mm (Figure 1) with average density ρ = 428.4 ± 27.9 kg·m^{-3}. Samples were cut tangentially by continuous laser powers 100 and 150 W, with cutting speeds of (3, 6, 9)·10^{-3} m·s^{-1}. The focal length was 1.27·10^{-1} m, beam diameter was 10^{-2} m, and spot diameter was 3·10^{-4} m. The focal point position of laser beam was set up to 1/2 of the sample thickness (measured from the upper surface of the board). The process gas was supplied through a Laval contour nozzle with compressed air of 0.25 MPa. The parameters of the cut were determined using digital microscopy through K-cluster analysis. The total number of measurements was 108 in one block. The ratio of kerf width on the upper and lower surface was calculated subsequently.

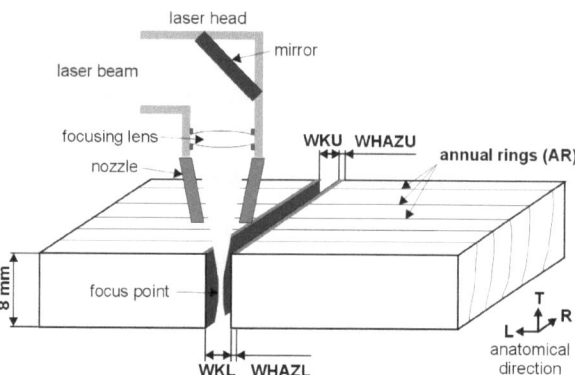

Figure 1. Cutting scheme.

Abbreviations are used in the text, the meaning of which is as follows: cutting kerf width on upper surface (WKU), cutting kerf width on lower surface (WKL), ratio of WKU and WKL (WKR), width of heat affected zone on upper surface (WHAZU), width of heat affected zone on lower surface (WHAZL), laser power (P) and number of annual rings (AR). Definitions of investigated parameters are described in Figure 1.

In the research, we define 5 cut characteristics whose meaning are as follows: cutting kerf width on upper surface (WKU), cutting kerf width on lower surface (WKL), ratio of WKU and WKL (WKR), width of heat affected zone on upper surface (WHAZU), width of heat affected zone on lower surface (WHAZL), laser power (P) and number of annual rings (AR).

Basics of Artificial Neural Networks

Artificial neural networks are used for prediction of materials properties when analytical mathematical approximation cannot be found. From this very robust mathematical tool, material properties can be predicted.

The advantages of neural networks are as follows:

- The neural network can learn
- The neural network can generalize

The disadvantages are as follows:

- Neural networks need more values of one or more parameters that change in every dataset as input compared with standard fitting procedures such as the-least squares method and many others.

ANN usage is tested by statistical parameters root mean square error RMSE, coefficient of determination R^2 and the slope between the predicted and real measured values and the relative root mean square error defined as the ratio of RMSE and minimal measured value of predicted the parameter.

3. Results and Discussion

The results and discussion section will be divided into three parts namely measuring the parameters results, ANN prediction results and studying the AR and P effect on WKU, WKL, WKR, WHAZU, WHAZL. All the results were discussed with literature references.

3.1. Measured Values

Table 1 shows the average values with standard deviation for all input values of P equal to 100 and 150 W. The figures in Table 2 show the average values together with the standard deviations for all the input values of v equal to $(3, 6, 9) \cdot 10^{-3}$ m·s^{-1}. In Table 3 the average values together with the standard deviations for all the input values of AR from 3 to 11 are shown.

Table 1. Descriptive statistics of all the measured parameters vs. P.

Value at P	WKU (10^{-3} m)	WKL (10^{-3} m)	WKR (-)	WHAZU (10^{-3} m)	WHAZL (10^{-3} m)
100	0.81 ± 0.17	0.50 ± 0.09	0.48 ± 0.04	0.16 ± 0.02	0.18 ± 0.02
150	0.91 ± 0.22	0.58 ± 0.14	0.77 ± 0.06	0.19 ± 0.02	0.26 ± 0.03

Table 2. Descriptive statistics of all the measured parameters vs. v.

Value at v	WKU (10^{-3} m)	WKL (10^{-3} m)	WKR (-)	WHAZU (10^{-3} m)	WHAZL (10^{-3} m)
3	1.04 ± 0.11	0.65 ± 0.08	0.64 ± 0.17	0.15 ± 0.02	0.20 ± 0.04
6	0.91 ± 0.13	0.56 ± 0.07	0.61 ± 0.16	0.18 ± 0.03	0.22 ± 0.06
9	0.62 ± 0.04	0.40 ± 0.02	0.63 ± 0.13	0.19 ± 0.02	0.23 ± 0.05

Table 3. Descriptive statistics of all the measured parameters vs. AR.

Value at AR	WKU (10^{-3} m)	WKL (10^{-3} m)	WKR (-)	WHAZU (10^{-3} m)	WHAZL (10^{-3} m)
3	0.90 ± 0.24	0.57 ± 0.15	0.67 ± 0.19	0.17 ± 0.03	0.21 ± 0.06
4	0.91 ± 0.23	0.57 ± 0.14	0.66 ± 0.19	0.17 ± 0.03	0.21 ± 0.06
5	0.90 ± 0.23	0.56 ± 0.15	0.64 ± 0.18	0.16 ± 0.03	0.21 ± 0.06
6	0.87 ± 0.23	0.54 ± 0.14	0.63 ± 0.16	0.17 ± 0.03	0.21 ± 0.06
7	0.85 ± 0.22	0.53 ± 0.13	0.62 ± 0.16	0.17 ± 0.03	0.22 ± 0.05
8	0.81 ± 0.19	0.51 ± 0.12	0.60 ± 0.15	0.17 ± 0.03	0.22 ± 0.05
9	0.83 ± 0.21	0.52 ± 0.13	0.61 ± 0.17	0.18 ± 0.03	0.22 ± 0.05
10	0.82 ± 0.18	0.51 ± 0.11	0.61 ± 0.14	0.18 ± 0.02	0.23 ± 0.06
11	0.82 ± 0.19	0.52 ± 0.11	0.61 ± 0.14	0.19 ± 0.02	0.23 ± 0.06

3.2. ANN Prediction

In this article, we have used values of laser power 100 and 150 W, the number of annual rings 3 to 11 and cutting speed $(3, 6, 9) \cdot 10^{-3}$ m·s^{-1} as input values for teaching and measured values of WKL, WKU, WHAZL and WHAZU and WKR, thus the training groups have 54 lines. By routine, we obtained the five best artificial neural networks for prediction of WKU, WHAZU, WKL, WHAZL and WKR.

In the measured values teaching and generalization we have used all the basic neural networks used in the Statistica 12 software. According to a characteristic sum of squares, we obtained the five best neural networks which are shown in Table 4. All the studied multilayer perceptron networks used the Quasi–Newton training algorithm and the best neural network MLP 3-3-5 used the BFGS 73 training algorithm with error function sum of squares. The hidden layer was activated by logistic function and the output activation function was exponential. The number of the hidden neurons' possibilities were between 1 and 54 (length of dataset).

Table 4. Error propagation in the best neural networks.

Net	Training Error (-)	Testing Error (-)	Validation Error (-)
MLP 3-7-5	0.019	0.023	0.022
MLP 3-3-5	0.010	0.013	0.011
MLP 3-10-5	0.018	0.022	0.022
MLP 3-6-5	0.004	0.014	0.011
MLP 3-10-5	0.030	0.030	0.036

In this article we will present the results of the P and AR effect on the abovementioned parameters of wood cutting for all the possible parameters of CO_2 lasers from which it is possible to predict Picea abies wood cutting properties at any laser power between 100 and 500 W and the number of annual rings between 3 and 11 with a goal to optimize the cutting process. The results are given for Picea abies wood, therefore they can be successfully used for CO_2 laser cutting of Picea abies.

Artificial neural networks have been used for prediction of WKU, WKL, WKR, WHAZU, WHAZL parameters for input parameters P = 200, 300, 400, 500 W, cutting speed $v = 1.2 \cdot 10^{-2}$ m·s^{-1} and the number of annual rings AR = 3, 4, 5, ..., 11. The error propagation of the networks are presented in Table 4. All the types of errors are decimal form of variance coefficient.

From Table 4, it is good to see that the lowest error (the standard is the mean square error) is obtained for network MLP 3-3-5 which was also found to be the best according to the slope of the predicted vs. measured values, coefficient of determination R^2, root mean square error RMSE and relative root mean square error Rel_RMSE, which are shown in Table 5.

Table 5. Statistical parameters of the ANN networks.

Statistical Parameter of	WKU (10^{-3} m)	WKL (10^{-3} m)	WKR (-)	WHAZU (10^{-3} m)	WHAZL (10^{-3} m)
Slope (-)	0.970	0.960	0.940	0.960	0.970
R^2 (-)	0.950	0.980	0.933	0.960	0.970
RMSE (parameter units)	0.010	0.004	0.012	0.001	0.002
Rel_RMSE (-)	0.018	0.012	0.035	0.009	0.013

From Table 5, we can conclude:

- The coefficient of determination R^2 for all the output parameters dependent on the input parameters is at least 0.933.
- The slope which should be close to 1 is for all the studied parameters higher than 0.94, which is a very good result.
- The maximum value of Rel_RMSE which in percentage is equal to 3.5% is highly under the measuring accuracy from which we can conclude that ANN can predict with accuracy all the studied cutting results' parameters.
- ANN can teach and predict with very high accuracy all the studied parameters.
- In the next three chapters we will deal with prediction of (WKU, WHAZU), (WKL, WHAZL) and WKR vs. P and AR and discuss the effects of P and AR on all the predicted parameters.

3.3. Prediction of WKU, WHAZU versus AR and P

In this chapter we will deal with prediction of WKU and WHAZU vs. AR, P at non-measured parameters which give information on the quality of Picea abies cut on the upper board. The results we will be presented as a graph of output property vs. AR, P in categories of laser power at a cutting speed of $1.2 \cdot 10^{-2}$ m·s^{-1}.

In Figure 2a,b we are show the dependence WKU and WHAZU vs. AR and P for cutting speed $1.2 \cdot 10^{-2}$ m·s^{-1} in this order.

From Figure 2a,b we can see that AR have effect on WKU only for laser power 200 W and don't have effect on WHAZU. WKU is changing in average from 0.64 mm to 0.81 mm which correspond to 26.5% which is above the measuring error of this dimension. Therefore, we can conclude that AR does have statistically significant effect on WKU. Change of WKU vs. AR at low power 200 W lies in fact that at 200 W is material removal rate MRR low thus heat transfer through higher thermal conductivity by higher AR is significant which leads to accumulation of heat in kerf region thus higher cutting kerf width which is proportional to mass of wood which is burned.

At higher laser powers WKU don't change with AR and P which can be described by fact that at these values of P material removal rate MRR is at maximal value and therefore heat transfer in the cutting kerf region don't occur. This fact was also recorded by many authors in literature like as Hernandez (16) on pine wood and Barnekov (13) and Asibu (6). From the Figure 2 it is also good to see that laser power increases the cutting kerf width at upper board which was found out also by Nukman (8), Ready (5) and Liu (18).

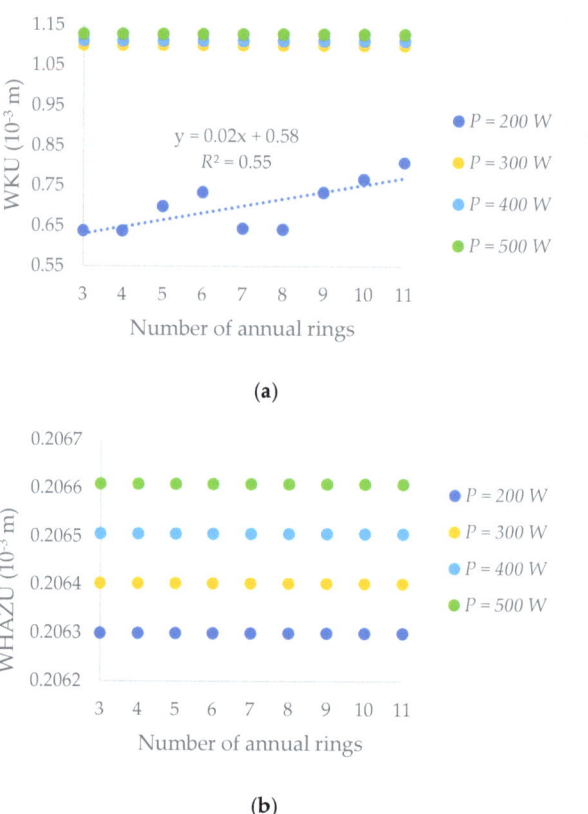

Figure 2. (a) WKU vs. AR, P at $1.2 \cdot 10^{-2}$ m·s^{-1}; (b) WHAZU vs. AR, P at $1.2 \cdot 10^{-2}$ m·s^{-1}.

From Figure 2b is good to see that laser power don't have significant effect on WHAZU which is caused by fact that at higher cutting speed upper board thickness is cut quicker and therefore heat propagation in HAZ region don't occur which is in good agreement with Asibu (6), Barcikowsky (9) and Lum (15). All presented results in Figure 2 is in good agreement with Kubovský (7).

3.4. Prediction of WKL, WHAZL versus AR, P

In this chapter we will deal with prediction of WKL and WHAZL vs. AR, P at non-measured input parameters which give information about quality of Picea abies cut on lower board. The results we will present as graph of output property vs. AR, P in categories of laser power at cutting speed $1.2 \cdot 10^{-2}$ m·s^{-1}.

In the Figure 3a,b we are showing dependence of WKL and WHAZL vs. AR and P for a cutting speed of $1.2 \cdot 10^{-2}$ m·s^{-1} in this order.

From Figure 3a,b we can see that AR has an effect on WKL only for a laser power pf 200 W and does not have an effect on WHAZL. WKL however changes on average from 0.6 mm to 0.53 mm which corresponds to 11.7% decrease. During cutting of wood HAZ increases with depth of cut and this is mainly because of local heat sources above the lower board which lead to heat transfer from the "higher" parts of the wood to the lower surface. This also leads to enlargement of the width of HAZ together with fact that at a lower board, the heat transfer to the heat affected zone is more significant thus the heat generated in the cutting kerf region is transferred to the heat affected zone. This fact also is in good agreement with Figures 2b and 3b from which it is good to see that the width of heat affected zone at the lower board 0.29 mm is statistically significantly higher than for the upper board of 0.21 mm which correspond to a 39% increase. However, a decrease of WKL vs. AR is on the bounds of measurement error.

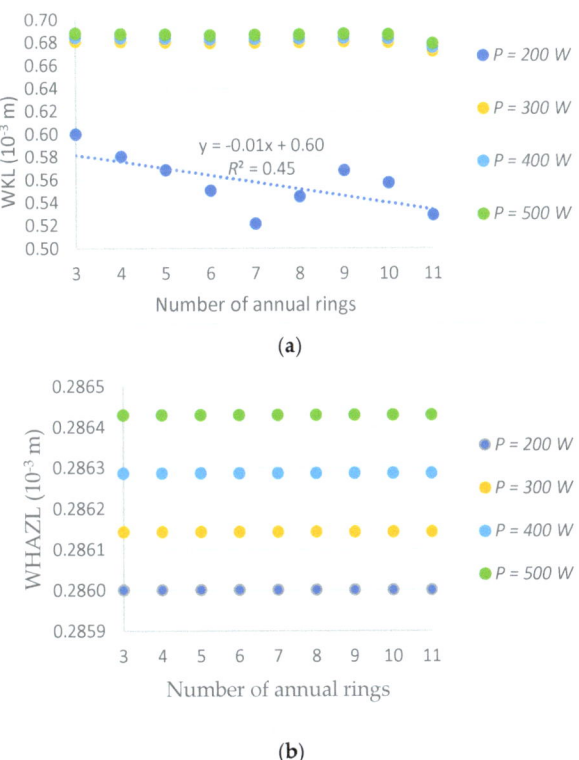

Figure 3. (a) WKL vs. AR, P at $1.2 \cdot 10^{-2}$ m·s^{-1}; (b) WHAZL vs. AR, P at $1.2 \cdot 10^{-2}$ m·s^{-1}.

On the other side, similar to the cutting kerf width at the upper board WKU, WKL increases with laser power only for laser powers lower than 300 W at which the cutting kerf width at the lower board reaches stabilization. This describes why the values of WKL have statistically equal values for all laser powers between 300 and 500 W. These results have been recorded in many scientific journal papers such as those by Ready (5), Lum (15), Lum (10), Eltawahni (11) and Kubovský (7).

From Figure 3b, it is good to see that P and AR do not have a significant effect on WHAZL. The reason for this lies in the fact that P and AR are not the main reasons for the width of heat affected zone increasing, which is caused by heat transfer from the upper parts of the wood to the lower parts of wood.

3.5. Comparison of P and AR Effect on Cutting Kerf Width and Width of the Heat Affected Zone for the Upper Board and Lower Board

From Figures 2 and 3, we can conclude:

- For all values of P, AR value of WKU is statistically significantly higher than for WKL which is caused by the fact that at lower board heat transfer, it plays a more significant role in the kerf region.

- For all values of P, AR value of WHAZU is statistically significantly lower than for WHAZL which is caused by the fact that at lower board heat transfer, it plays a more significant role in the heat affected zone region.
- The smallest difference between the values of WKU and WKL is at a minimal power of 200 W.
- AR and P do not play a significant role in both WHAZU and WHAZL.
- WHAZU is significantly lower than WHAZL because of heat transfer from the upper board to the lower board in the HAZ region.

3.6. Prediction of WKR versus Number of Annual Rings AR and Laser Power P

In this chapter we will deal with the prediction of WKR vs. AR, P at non-measured input parameters which give information on the quality of Picea abies cut at whole thickness. The results we will present as presented in a graph of output property vs. AR, P in categories of laser power at a cutting speed of $1.2 \cdot 10^{-2}$ m·s^{-1}.

In Figure 4, we show the dependence of WKR vs. the AR and P for a cutting speed of $1.2 \cdot 10^{-2}$ m·s^{-1}.

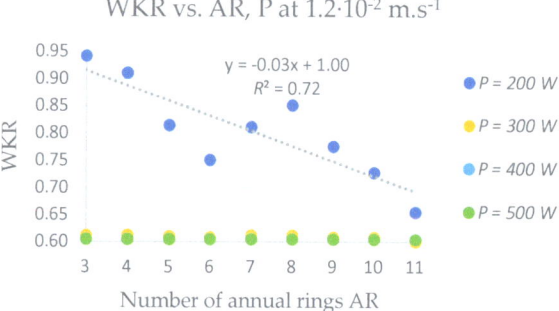

Figure 4. WKR vs. AR, P at $1.2 \cdot 10^{-2}$ m·s^{-1}.

From Figure 4, we can see that AR has an effect on WKR only for a laser power of 200 W. WKR decreases with AR cca. 3% by increased values of AR by 1 thus statistically significant changes occur only for values of AR = 8 and higher. From this graph, it is good to see that increasing P causes worse cutting therefore it is better to use lower laser values at a cutting velocity of $1.2 \cdot 10^{-2}$ m·s^{-1}. Changes of WKR at higher values of AR are caused by the fact that with increasing AR, heat transfer through thermal conductivity of wood plays a more significant role at the lower board vs. the upper board. The lower value of laser power is also good according to manufacturing costs. The presented results are in good agreement with Eltawahni (11) and Kubovský (7).

4. Conclusions

The results of this study suggest that the following conclusions:
- According to Tables 4 and 5, we can conclude that artificial neural networks are able to predict values of all the studied parameters of cutting, with MLP 3-3-5 which has low error backpropagation characteristics being the optimal neural network.
- The width of a kerf at the upper surface (WKU) and the width of a kerf at the lower surface (WKL) do not increase with a P higher than 300 W.
- Both AR and P do not have a significant effect on WHAZU and WHAZL.
- AR changes WKR only when P is lower than 300 W, thus AR has an effect on WKR only for low CO_2 laser powers.

- The values of WKU for all the studied parameters of P and AR are statistically significantly higher than for WKL which is in contrast with the increased value of WHAZL vs. WHAZU at cca. 40% level.
- The values of WKR at a P higher than 200 W are cca. equal to 0.7 which is a very low value therefore higher powers lead to worse cutting quality so we can conclude that they are not appropriate for a cutting speed of $1.2 \cdot 10^{-2}$ m·s^{-1}.
- Generally, the value of P = 200 W improves the cutting quality in the terms of WKR values and therefore they are optimal for wood cutting at $1.2 \cdot 10^{-2}$ m·s^{-1}. The values of WKR at P = 200 W are higher than the values of this parameter at 100 and 150 W even if the cutting speed at optimization ($1.2 \cdot 10^{-2}$ m·s^{-1}) is higher than at the measured dataset (max $0.9 \cdot 10^{-2}$ m·s^{-1}).
- Prediction of the cutting conditions of spruce wood machined by CO_2 laser effects on cut characteristics, obtained by artificial neural network MLP 3-3-5, can be used for qualitative characterization of the cutting conditions in wooden technological processes.
- In our article, we predicted the effect of P and AR by constant value of v on the cut characteristics and therefore we used regression modelling using ANN in Statistica software. According to abovementioned parameters this showed a very good correlation, and a low sum of squares error, which for characterization of wood materials cutting is highly under the error level at which these properties are measured, which is in good agreement with other authors' results.
- In the next stage of research, we will focus on studying the cutting speed v, the number of annual circles (AR) and the laser power P effect on all above-mentioned parameters, as it is very useful to compare the results with other approaches, which should be deep learning approaches, non-linear parametric fitting procedures and many others to optimize the accuracy of the prediction of cut characteristics.

Author Contributions: Conceptualization, I.R. and R.I.; methodology, R.I. and I.R.; software, I.R.; validation, I.R., R.I. and I.K.; formal analysis, I.K. and M.G. and A.J.; investigation, I.R., R.I. and I.K.; resources, I.K.; data curation, I.R. and R.I.; writing—original draft preparation, I.R.; writing—review and editing, R.I. and I.K.; visualization, A.J.; supervision, I.K.; project administration, I.K.; funding acquisition, I.K. All authors have read and agreed to the published version of the manuscript.

Funding: This work was supported by the Slovak Research and Development Agency under the Contract no. APVV-20-0159 (80%) and by the VEGA Agency of the Ministry of Education, Science, Research, and Sport of the Slovak Republic and the Slovak Academy of Sciences Grant no. 1/0577/22 (20%).

Institutional Review Board Statement: Not applicable.

Informed Consent Statement: Not applicable.

Data Availability Statement: Not applicable.

Conflicts of Interest: The authors declare no conflict of interest. The funders had no role in the design of the study; in the collection, analyses, or interpretation of data; in the writing of the manuscript, and in the decision to publish the results.

References

1. Abidou, D.; Yusoff, N.; Nazri, N.; Awang, M.A.O.; Hassan, M.A.; Sarhan, A.A.D. Numerical simulation of metal removal in laser drilling using radial point interpolation method. *Eng. Anal. Bound. Elem.* **2017**, *77*, 89–96. [CrossRef]
2. Mukherjee, K.T.; Grendzwell, P.A.A.; McMillin, C.W. Gas-flow parameters in laser cutting of wood-nozzle design. *For. Prod. J.* **1990**, *40*, 39–42. Available online: https://www.fs.usda.gov/research/treesearch/8245 (accessed on 11 September 2022).
3. Sinn, G.; Chuchala, D.; Orlowski, K.; Tauble, P. Cutting model parameters from frame sawing of natural and impregnated Scots Pine (*Pinus sylvestris* L.). *Eur. J. Wood Wood Prod.* **2020**, *78*, 777–784. [CrossRef]
4. Martinez-Conde, A.; Krenke, T.; Frybort, S.; Muller, U. Review: Comparative analysis of CO_2 laser and conventional sawing for cutting of lumber and wood-based materials. *Wood Sci. Technol.* **2017**, *51*, 943–966. [CrossRef]
5. Ready, J.F.; Farson, D.F.; Feeley, T. *LIA Handbook of Laser Materials Processing*; Laser Institute of America, Magnolia Publishing: Orlando, FL, USA; Springer Nature: Berlin/Heidelberg, Germany, 2001.

6. Asibu, E.K. *Principles of Laser Materials Processing*; John Wiley and Sons: New York, NY, USA, 2009.
7. Kubovský, I.; Krišťák, Ľ.; Suja, J.; Gajtanska, M.; Igaz, R.; Ružiak, I.; Réh, R. Optimization of Parameters for the Cutting of Wood-Based Materials by a CO_2 Laser. *Appl. Sci.* **2020**, *10*, 8113. [CrossRef]
8. Nukman, Y.; Saiful, R.I.; Azuddin, M.; Aznijar, A.Y. Selected Malaysian Wood CO_2 Laser Cutting Parameters and Cut Quality. *Am. J. Appl. Sci.* **2008**, *5*, 990–996. [CrossRef]
9. Barcikowski, S.; Koch, G.; Odermatt, J. Characterisation and modification of the heat affected zone during laser material processing of wood and wood composites. *Holz Roh Werkst.* **2006**, *64*, 94–103. [CrossRef]
10. Lum, K.C.P.; Hg, S.L.; Black, I. CO_2 laser cutting of MDF: Determination of process parameter settings. *J. Opt. Laser Technol.* **2000**, *32*, 67–76. [CrossRef]
11. Eltawahni, H.A.; Olabi, A.G.; Benyounis, K.Y. Investigating the CO_2 laser cutting parameters of MDF wood composite material. *Opt. Laser Technol.* **2011**, *43*, 648–659. [CrossRef]
12. Eltawahni, H.A.; Rossini, N.S.; Dassisti, M.; Alrashed, K.; Aldaham, T.A.; Benyounis, K.Y.; Olabi, A.G. Evaluation and optimization of laser cutting parameters for plywood materials. *Opt. Lasers Eng.* **2013**, *51*, 1029–1043. [CrossRef]
13. Barnekov, V.G.; McMillin, C.W.; Huber, H.A. Factors influencing laser cutting of wood. *For. Prod. J.* **1986**, *36*, 55–58.
14. Barnekov, V.G.; Huber, H.A.; McMillin, C.W. Laser machining wood composites. *For. Prod. J.* **1989**, *39*, 76–78. [CrossRef]
15. Ng, S.L.; Lum KC, P.; Black, I. CO_2 laser cutting of MDF:2. Estimation of power distribution. *J. Opt. Laser Technol.* **2000**, *32*, 77–87. [CrossRef]
16. Hernandez-Castaneda, C.J.; Sezer, K.H.; Li, L. The effect of moisture content in fibre laser cutting of pine wood. *Opt. Lasers Eng.* **2011**, *49*, 1139–1152. [CrossRef]
17. Riveiro, A.; Quintero, F.; Boutinguiza, M.; del Val., J.; Comesana, R.; Lusquinos, F.; Pou, J. Laser Cutting: A Review on the Influence of Assist Gas. *Materials* **2019**, *12*, 157. [CrossRef]
18. Liu, Q.; Yang, C.; Xue, B.; Miao, Q.; Liu, J. Processing Technology and Experimental Analysis of Gas-assisted Laser Cut Micro Thin Wood. *BioResources* **2020**, *15*, 5366–5378. [CrossRef]
19. Dhanachandra, N.; Manglem, K.; Chanu, Y.J. Image Segmentation Using K-means Clustering Algorithm and Subtractive Clustering Algorithm. *Procedia Comput. Sci.* **2015**, *54*, 764–771. [CrossRef]
20. Fukuta, S.; Nomura, M.; Ikeda, T.; Yoshizawa, M.; Yamasaki, M.; Sasaki, Y. UV laser machining of wood. *Eur. J. Wood Wood Prod.* **2016**, *74*, 261–267. [CrossRef]
21. Fukuta, S.; Nomura, M.; Ikeda, T.; Yoshizawa, M.; Yamasaki, M.; Sasaki, Y. Wavelength dependence of machining performance in UV-, VIS- and NIR-laser cutting of wood. *J. Wood Sci.* **2016**, *62*, 316–323. [CrossRef]
22. Kúdela, J.; Kubovský, I.; Andrejko, M. Surface properties of beech wood after CO2 laser engraving. *Coatings* **2020**, *10*, 77. [CrossRef]
23. Kúdela, J.; Kubovský, I.; Andrejko, M. Impact of different radiation forms on beech wood discolouration. *Wood Res.* **2018**, *63*, 923–934. Available online: http://www.woodresearch.sk/wr/201806/01.pdf (accessed on 30 August 2022).
24. Kubovský, I.; Kačík, F.; Veľková, V. The effects of CO_2 laser irradiation on color and major chemical component changes in hardwoods. *BioResources* **2018**, *13*, 2515–2529. [CrossRef]
25. Kubovský, I.; Kačík, F.; Reinprecht, L. The impact of UV radiation on the change of colour and composition of the surface of lime wood treated with CO_2 laser. *J. Photochem. Photobiol. A Chem.* **2016**, *322*, 60–66. [CrossRef]
26. Avramidis, S.; Liadis, L. Predicting wood thermal conductivity using artificial neural networks. *Wood Fiber. Sci.* **2005**, *37*, 682–690. Available online: https://wfs.swst.org/index.php/wfs/article/view/260/260 (accessed on 15 August 2022).
27. Tiryaki, S.; Malkocoglu, A.; Ozsahin, S. Using artificial neural networks for modeling surface roughness of wood in machining process. *Constr. Build. Mater.* **2014**, *66*, 329–335. [CrossRef]
28. Tiriyaki, S.; Ozsahin, S.; Aydin, A. Employing artificial neural networks for minimizing surface roughness and power consumption in abrasive machining of wood. *Eur. J. Wood Wood Prod.* **2017**, *75*, 347–358. [CrossRef]
29. Tiryaki, S.; Malkocoglu, A.; Ozsahin, S. Artificial neural network modelling to predict optimum power consumption in wood machining. *Drewno* **2016**, *59*, 109–125. [CrossRef]
30. Akyuz, I.; Ozsahin, S.; Tiryaki, S.; Aydin, A. An application of artificial neural networks for modelling formaldehyde emission based on process parameters in particleboard manufacturing process. *Clean Technol. Environ. Pol.* **2017**, *19*, 1449–1458. [CrossRef]
31. Tiryaki, S.; Aras, U.; Kalaycioglu, H.; Erisir, E.; Aydin, A. Predictive Models for Modulus of Rupture and Modulus of Elasticity of Particleboard Manufactured in Different Pressing Conditions. *High Temp. Mat. Proc.* **2017**, *36*, 623–634. [CrossRef]
32. Bardak, S.; Tiryaki, S.; Bardak, T.; Aydin, A. Predictive Performance of Artificial Neural Network and Multiple Linear Regression Models in Predicting Adhesive Bonding Strength of Wood. *Strength Mater.* **2016**, *48*, 811–824. [CrossRef]
33. Akyuz, I.; Ersen, N.; Tiryaki, S.; Bayram, B.C.; Akyuz, K.C.; Peker, H. Modelling and comparison of bonding strength of impregnated wood material by using different methods: Artificial neural network and multiple linear regression. *Wood Res.* **2019**, *64*, 483–497. Available online: http://www.woodresearch.sk/.../10.pdf (accessed on 29 September 2022).
34. Bardak, S.; Tiryaki, S.; Nemli, G.; Aydin, A. Investigation and neural network prediction of wood bonding quality based on pressing conditions. *Int. J. Adh. Adh.* **2016**, *68*, 115–123. [CrossRef]
35. Nguyen, T.T.; Nguyen, T.H.V.; Ji, X.; Yuan, B.; Trinh, H.M.; Do, K.T.L.; Guo, M. Prediction of the color change of heat-treated wood during artificial weathering by artificial neural network. *Eur. J. Wood Wood Prod.* **2019**, *77*, 1107–1116. [CrossRef]

36. Nguyen, T.H.V.; Nguyen, T.T.; Ji, X.; Guo, M. Predicting color change in Wood During Heat Treatment using an artificial neural network model. *BioResources* **2018**, *13*, 6250–6264. [CrossRef]
37. Demir, A.; Cakiroglu, E.O.; Aydin, I. Determination of CNC processing parameters for the best wood surface quality via artificial neural network. *Wood Mater. Sci. Eng.* **2021**. [CrossRef]
38. Demir, A.; Birinci, A.U.; Ozturk, H. Determination of the surface characteristics of medium density fibreboard processed with CNC machine and optimisation of CNC process parameters by using artificial neural networks. *J. Manuf. Sci. Technol.* **2021**, *35*, 929–942. [CrossRef]
39. Dou, Z.; Sun, Y.; Zhang, Y.; Wang, T.; Wu, C.; Fan, S. Regional Manufacturing Industry Demand Forecasting: A Deep Learning Approach. *Appl. Sci.* **2021**, *11*, 6199. [CrossRef]

Surface Engineering of Woodworking Tools, a Review

Bogdan Warcholinski * and Adam Gilewicz

Faculty of Mechanical Engineering, Koszalin University of Technology, 75-453 Koszalin, Poland
* Correspondence: bogdan.warcholinski@tu.koszalin.pl

Abstract: The wide range of applications of wood are due to its strength properties. The mechanical properties of wood in various parts or directions are different. The complex structure of wood and its hygroscopicity prevent the use of coolants and lubricants, resulting in rapid tool wear disproportionate to the hardness of the processed material. This significantly affects machining efficiency and the quality of the processed surface. It seems that an effective method of reducing tool wear is its modification with a thin hard coating produced by the Physical Vapor Deposition or Chemical Vapor Deposition methods. The article presents tool materials used for woodworking, areas for improving the efficiency of their work, and the impact of thin hard coatings on the increase in tool durability, including binary coatings and also doping with various elements and multilayer coatings. Scientific centers dealing with the above-mentioned subject are also mentioned. A brief review of the effects of surface modifications of woodworking tools in the context of their durability is presented. It was found that the most promising coatings on tools for woodworking were multilayer coatings, especially based on chromium. Higher wear resistance was demonstrated by coatings with a lower coefficient of friction. This value was more important than hardness in predicting the service life of the coated tool.

Keywords: tool modification; woodworking tool; lifetime of the tool

1. Introduction

About 29% of the Earth's surface and over 44% of Europe is covered by forests. European forests represent about 25% of the world's forest resources. Forested areas have still been increasing, for the past 20 years by about 0.8 million ha per year.

Wood is an important raw material in construction, furniture, heating, the packaging industry, etc. The huge popularity of wood as a raw material results from its special properties and aesthetic qualities. It is an excellent, healthy, and renewable material with a complex structure consisting of many types of cells and substances present in different amounts, depending on the species of wood and its parts. In order to reduce unit manufacturing costs, woodworking plants and furniture companies strive to increase production of manufactured goods. Investments in modern tools, but also modern systems for processing wood and wood-based materials, devices for assessing wood quality, and processing technologies make it possible to meet growing environmental requirements and reduce production costs.

The key factor influencing the effectiveness of wood and wood-based product processing is the life of tools. This is directly related to tool wear and indirectly affects power consumption and the surface quality of the workpieces. The use of advanced high-performance cutting tools allows reduction in operating costs, which is directly responsible for increasing the productivity of the technology used. One of the factors is the selection of the tool material and the condition of machines. Processing hardwood species (oak, ash, hornbeam, ebony, pink lapacho) may require tools made of different materials than for soft wood (spruce, pine, larch) processing. The structure and chemical composition of wood species, including the amounts of minerals and resin, significantly affect this choice. For example, according to Kadur company [1], stellite planer knives are suitable for working with hard and semi-hard wood, and HSS planer knives of alloyed tungsten (6–18%) are suitable for semi-hard and soft woods.

In the wood industry, tungsten carbide tools are widely used for processing fiberboards, chipboards, and solid wood, as the successor to high-speed steel tools. However, they also show a relatively high degree of wear in some applications. Although they prove superior to high speed steel tools in many applications, the use of tungsten carbide tools in some particle board and fiberboard applications is limited due to the relatively high degree of wear. Therefore, there is an urgent need to search for new materials and technologies for their production that would improve the wear resistance of tools.

The analysis of the literature shows that, after the initial failure to improve the effectiveness of tools by applying a protective coating, current data indicate a significant improvement in abrasion resistance with a properly prepared substrate for the coating. One of the most commonly used methods to increase tool life is to apply thin, hard coatings to the cutting tool using physical vapor deposition (PVD) or chemical vapor deposition (CVD).

There are many reputable journals covering wood materials and wood-based products, the biology and physics of wood, wood processing technologies, and the application of these materials. These are, i.a., *Bioresources, European Journal of Wood and Wood Products, Journal of Wood Chemistry and Technology, Wood Material Science & Engineering, Wood Science and Technology*, and others. In many journals, there are special issues devoted to this subject; for example, in *Applied Sciences* there is *Application of Wood Composites, Advances in Wood processing Technology* [2–9].

The article presents a brief outline of the use of selected surface modifications of tools for processing wood and wood-like materials. The aim of the article is the chronological documentation of research conducted in this area, taking into account the type of wood processing and the tool used for it, and not the performance of a comprehensive analysis of the literature on wood processing. Due to the large differences in the structures of metal and wood (cellular nature, anisotropy, and multi-scale level organization), the test results of metalworking tools with a modified (hard thin coating) working surface cannot be directly applied to woodworking tools.

2. Tool materials

Due to machining conditions, cutting tools should be characterized by high resistance to mechanical loads and high temperatures. In the case of metal processing, frictional heat can increase the temperature in the cutting zone to 700 °C [10], but during wood processing and especially the processing of wood-like materials, the temperature exceeds 800 °C, which results from high friction forces in the machining zone [11]. Such temperature occurs both on the rake surface and the clearance surface, and has a significant impact on the wear rate of the cutting tool. The main parameters which ought to be taken into account are: chemical inertness and stable physical properties, including hardness at high temperatures, low wear ratio for different wear mechanisms, and sufficient toughness to avoid material fracture [12].

The history of the development of tool materials used for wood (wood-like materials) and metal processing is similar. Tool steels were applied as the first materials; they were relatively soon replaced by high-speed steels. The increased interest in wood-based material processing and related problems resulted in the introduction of composite tools based on hard, fine WC particles sintered with cobalt. Co is responsible for the elastic bounding of hard WC particles. By selecting the size of hard particles and the amount of bonding phase, the mechanical properties of sintered tools can be changed. Another group of tool materials is stellites, characterized by high hardness depending on the chemical composition (36–52 HRC), and high resistance to abrasion and high temperatures (up to 950 °C).

In the early 1980s, one of the most modern tools was sintered carbide. After the dramatic question "Is there life after tungsten carbide?" [13], the answer was almost immediate, with "PCD replacing carbide in woodworking applications" [14]. At this time, there was an abrupt increase in interest in modified woodworking tools. The polycrystalline diamond (PCD) displaced traditional steel and cemented carbide tools in woodworking [15–18]. The significantly better strength and tribological properties of PCD, such as hardness (about 40–50 GPa, compared to 10–22 GPa for carbides) and low coefficient of friction, meant it could enable an increase in durability and minimize the machine downtime of tools. This meant that PCD tools offered considerable potential for cost savings in the machining of wood and wood-like materials. Boyle indicated [15] that the life ratio of PCD tools compared to carbide cutters was about 17:1. Only its price forced the search for new solutions. These were the thin hard coatings deposited on tools' working surfaces.

The application of tools with modified hard coatings for metalworking resulted in an abrupt increase in production capabilities and product quality. Knowledge of the research and exploitation of the coatings for metalworking tools cannot be fully adopted in woodworking. This is related to the differences in their structures and properties, including mechanical properties, coefficients of thermal conductivity, roughness, and absorption of various substances. This results in the need to design new coatings that meet market expectations.

Due to the hygroscopicity of wood and wood-like materials, an application of cooling and lubricating agents is excluded. It seems that the best perspective to adopt is the improvement of tool durability by surface modification. Klamecki [19], in one of the first literature reviews of wood cutting tool wear, indicated the problems with the measurement of edge dulling. He stated that tool wear deals primarily with the tool material, with the work material, and with tool–work interactions, and the chemical nature of wood may play a large part in cutting tool wear. Thibault et al. [20] summarized wood machining over the last 50 years in France. They found that tool wear depends on the cutting process parameters, and kind and quality of timber. Coating technologies may improve tool life. Based on the literature, they indicated steel tool nitriding [21–23], and thin coating with a hard material (Cr_xN_y type) using physical vapor deposition gave promising results [24].

3. The Areas for Improving the Efficiency of Work Tools

The current requirements of users are mainly related to higher durability and reliability of tools and productivity. They expect an increase in the speed of machining, the ability of tools to work in automatic machining centers, and the processing of new, often difficult to process materials. Therefore, investigations are needed to design, manufacture, test on a laboratory scale, and implement such tools with the above requirements. These expectations can be met through three groups of activities:

- Introduction of new materials for their production or modification of the properties of the materials used, and selection of the geometry of the working parts of the tools [13,20,25–37];
- Application of appropriate cooling and lubricating agents;
- Selection of the proper surface-coating system, and shaping of the surface properties in terms of increasing the durability of the tools [14,24,38–41].

In recent years, there has been a growing interest in the subject of woodworking tools. Figure 1 shows the number of publications on woodworking and woodworking tools as well as tools with working surfaces modified by thin hard coatings.

In the last few years, a systematic increase in the number of publications has been noticeable. The number of articles on uncoated tools is approximately four times that of articles on coated tools. These articles discuss the results of new coatings formed on a wide range of tools, and the processing of various materials under carefully selected conditions. The aim of many works is to correlate the type of coating with the processed material and processing conditions.

The analysis was made on the basis of the SCOPUS database, a scientific database maintained by the Elsevier publishing house, containing information about published scientific papers. The data presented in the Figure present the database resources, taking into account the terms "woodworking", "woodworking tool" and "woodworking tool + coating " searched within all fields in the database.

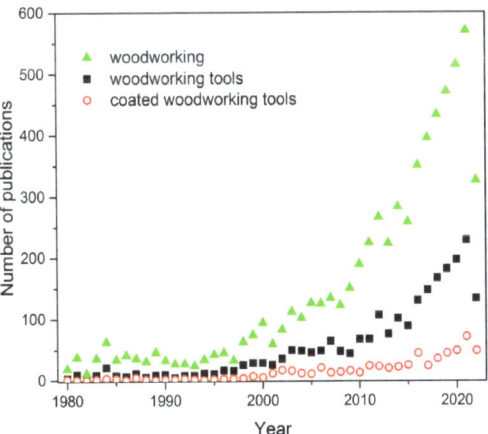

Figure 1. The subject of woodworking and woodworking tools in scientific articles, 1980–2022.

4. Surface Technologies

The development of tool materials looks to combine high hardness with high toughness, to improve their mechanical, tribological and thermophysical properties, leading to an improvement in the durability of cutting tools. One of the most important properties of the coating is good adhesion to the substrate and between the substrate layers. Other important properties are chemical stability and high wear resistance in working conditions. One can specify two groups of technologies that improve the wear resistance of tools.

The first one includes methods for improving the mechanical properties of the material tools (such as heat or thermo-chemical treatment). Heat treatment is to give them the desired mechanical properties by changing their structure, while the thermo-chemical treatment is to enrich the surface layer of the alloy in a certain element, such as C, N, Al, Cr, Si, or group of elements, e.g., C and N, N and S, N and O. The aim of these treatments is to give the surface layer specific physical properties—mainly resistance to abrasion, or chemicals—usually by resistance to oxidation at high temperatures. The quality of the tool is improved by applying heat treatment that allows it to obtain the desired hardness of the blade, and the appropriate fine-grained structure of steel and toughness.

Thermo-chemical treatment, especially nitriding, has a beneficial effect on the performance characteristics of tools. Many scientists indicate an increase in the wear resistance of nitrided tools for woodworking. Dependent on the type of technology used, this increase is as high as 100% [14,23,39].

The second group of technologies includes working surface modification techniques for tools by applying coatings with special properties. Among these, CVD (chemical vapor deposition) [42] and PVD (physical vapor deposition) [43] methods and surfacing using a submerged arc welding (SAW) technique and a mixture of alloying elements spread on the surface under industrial flux [44] can be listed.

The properties of the coatings can be modified by the type and roughness of the substrate, the deposition temperature, the composition and pressure of the gaseous atmosphere in the vacuum chamber, substrate bias voltage, and arc current or magnetron power, dependent on the deposition method. The above parameters influence the chemical and phase composition of the coatings, crystallite size, density of the coating, and surface quality (roughness). As an effect of properties such as hardness, elastic modulus, and also adhesion, toughness and wear resistance can be improved. The findings of Valleti et al. [45]

indicate that increase in substrate hardness results in higher adhesion of the coatings, but increase in substrate roughness results in lower adhesion of the coating investigated.

5. Coatings Used for Woodworking Tools

5.1. Binary Coatings

The first coatings applied were two-element systems. There are three main elements to synthesize protective coatings deposited on tools for woodworking: titanium [29,46–49], chromium [14,24,39,50–54] and carbon [39,49,53,55–57]. The transition metal nitrides were the most promising solution and they did not disappoint. One of the first tests in wood machining by TiN coating was sawing the following materials: hardboard, polyvinylchloride coated particleboard, waste paper-based paperboard, plywood, and spruce. The coatings were deposited by reactive triode ion plating on tungsten carbide with the thickness varied from 0.7 to 1.0 µm. The results indicated various effects on tool wear. On the one hand, when sawing hardboard and spruce, the coating reduced the wear on the rake face about 50% and 20%, respectively, compared with the uncoated tools. On the other hand, during sawing of particleboard, paperboard, and plywood, there was no increase in the durability of tools with a TiN coating [46]. This was one of the first signals that there is no universal coating. A proper coating should be selected for each treatment and type of workpiece. Additionally, other studies indicate that tool coatings can significantly improve wear resistance, especially when proper substrate preparation and proper tool geometry and cutting conditions are selected [29].

The advantages associated with the application of coatings to cutting tools are mainly high surface hardness, improved abrasion resistance, chemical inertness, and a relatively low coefficient of friction. After TiN coatings improving the durability of tools and machine parts, but also limiting applications and environment, attention was paid to chromium nitride coatings. These coatings, compared to coatings of other transition metal nitrides, are characterized by high resistance to corrosion and oxidation. Other features include high hardness and wear resistance, good adhesion of the coating to the steel substrate, high operating temperature, and a relatively low coefficient of friction [58–62]. A thin coating with a thickness of 2–3 µm may constitute a thermal barrier limiting heat transfer to the substrate. The results of investigations conducted by Kusiak et al. [50] indicated a significant reduction in the heat flux passing through the CrN coating, compared to the TiN coating. This was due to the nearly three times higher thermal conductivity of titanium nitride compared to chromium nitride. Single-layer CrN coatings are widely described in the literature both in the field of basic research and in applications, including woodworking. The higher resistance of the coatings to oxidation, and higher operating temperature are characterized by multi-component coatings in which one of the components is aluminum, e.g., TiAlN, forming thermally stable oxide coatings. In the cutting area, as a result of the forces acting on the blade, the oxide coating is removed and immediately, under the influence of high temperature, it is rebuilt.

It would seem that the thicker the coating on the tool, the longer it will last. Unfortunately, this is not true. Research carried out by Wiklund et al. [63] on a group of coating-substrate systems used in mechanical applications showed that thinner coatings are less susceptible to damage due to residual stresses as a result of adapting the geometric conditions at the interface. They also found that there is a critical coating thickness at which the stresses generated in it can cause the coatings to delaminate.

Durability tests of WC-Co inserts with CrN and Cr_2N coatings (thickness up to about 6 µm) synthesized by magnetron sputtering have shown that knives with coatings 1–2 µm thick have the best anti-wear properties. The service life of such tools was about four times higher than that of uncoated tools [24]. The thicker coatings showed signs of wear much earlier, probably due to the higher stresses in the coatings, confirmed by investigations conducted by Djouadi et al. [26].

The investigations conducted by Beer et al. [53] showed that CrN coatings on the 60SMD8 steel substrate are characterized by a smaller (about 52%) reduction of the knife edge compared to steel knives. W-C:H (DLC with WC precipitations) coatings show worse anti-wear properties. During the test, they showed delamination, and until the test was stopped, they showed only about 40% improvement in knife edge reduction. It should also be noted that the results varied significantly depending on the conditions of coating deposition and the preparation of the tool for the test. Also, during the test, decreasing power consumption and vibrations were observed for tools with CrN coatings, and the cutting process was stable [53]. Similar results are presented in [26]. Such good results for CrN coatings may have resulted from better adhesion of the coatings and a lower coefficient of friction, which also reduced the cutting forces and made the cutting process more stable in terms of vibration and veneer thickness [26].

5.2. Binary Coatings Doped with Various Elements

New requirements for tools, related to the increase in machining parameters or type of machining and the processing of new materials, led to the design and manufacture of new coatings. Two directions of change in coatings for woodworking tools were observed: formation of three- and more-component coatings, and forming of multilayer coatings. The influence of such elements as C, Si, Al, B, W, and Zr on the properties of coatings based on titanium [26,29,41,64,65] and chromium [38,41,66,67] was investigated. The next group, multilayer coatings, was also intensively studied [39,41,68–73]. Investigations of multilayered structures with chromium nitride as one of the layers indicated that it was possible to further improve hardness and toughness [14,74,75]. Some information about the applied coatings, the effects of their use, and the materials machined are summarized in Tables A1–A3. The coatings deposited on the tools are manufactured generally using PVD and CVD methods. Here, the results of PVD coatings are analyzed and summarized in Table A1 (magnetron sputtering) and Table A2 (cathodic arc evaporation). In Table A3, are gathered coatings formed and described by authors as Physical Vapor Deposition (PVD).

It should be noted that the first works including the assessment of the durability of the coating-modified tools did not reveal positive test results [76,77]. Titanium nitride (TiN), titanium carbide (TiC), and titanium aluminum oxynitride (TiAlON) were formed by plasma-assisted chemical vapor deposition on tungsten carbide tools applied to milling laminated particleboard. It was found that the tools with TiC provided only a slight improvement, and the tools with TiN and TiAlON did not provide any improvement in wear resistance compared to the uncoated carbide tools [77]. The analysis of the results of tests of TiN, TiAlN and TiN/TiCN coatings deposited by the PVD method on K grade tungsten carbide tools used for continuous milling of particleboards showed that wear resistance depends on the quality of the substrate, with the grain size and binder (cobalt) content being the decisive factors in the tool. For tools with a low cobalt content and small grains, there was a slight improvement in wear resistance, while coatings deposited on tools with a higher cobalt content and larger grains reduced the wear resistance. The authors observed the chipping of the coatings on the tool rake face, which was related with poor adhesion of the coating to the substrate (tool), and indicated this as the primary cause of poor wear resistance. [78]. Also, Darmawan et al. [68] indicated the poor adhesion of the coatings as the reason for its low wear resistance. For all tested systems: tool TiN, CrN, CrC, TiCN, and TiAlN, coating at both low and high cutting speeds of the wood-chip cement board, delamination occurred, as well as oxidation of the coating accelerated by the increase in cutting temperature. The above studies show the importance of adhesion to their wear resistance.

The research indicates that the most important factors determining the improvement of the performance of woodworking tools, but also of other types of processing, are good adhesion of the coating to the substrate (tool), low friction coefficient, and good abrasion resistance [26]. Many research centers have been involved in the optimization of these factors, and the results of their work can be found in many reputable journals. The final results of these works, with reference to specific applications for woodworking tools, are also included in this study. Benlatreche et al. [38] investigated CrN-based coatings with the addition of Al or Si, because such three-component coatings have modified structural, mechanical and tribological properties, including higher resistance to oxidation, higher hardness, and lower friction coefficient. Medium density fiberboard was subjected to routing tests. It has been found that the CrN coating improves the life of the cemented carbide tool by about 25 to 40%, depending on the coating deposition parameters and the deposition system used. The CrAlN coating had larger nose widths than the CrN coating, but with the increase in aluminum concentration in the coating, the difference in nose width between the coated and uncoated tool decreased. Depending on the type of MDF material being processed, the reduction in nose width was from about 25% (MDF fireproof) to about 40% (MDF standard). CrSiN coatings are characterized by slightly longer durability than CrN coatings. The nose width is reduced by approximately 35% compared to the uncoated tool and approximately 25% for the CrN coating. As for the aluminum coating, an increase in silicon concentration causes a decrease in its wear resistance. Above results are comparable with Ref. [69].

An important problem already at the early stage of tool testing was the monitoring and evaluation of the machining efficiency of woodworking tools [3,79,80].

5.3. Multilayer Coatings

Multilayer coatings generally have better wear resistance compared to single-layer coatings. This takes into account many factors such as: total thickness of the coating and individual layers, type of layer, functions performed in the coating (adhesive, abrasive, sliding, etc.). Greater wear resistance is connected with reduction in crack propagation at grain boundaries and layers. The coated carbide tools provide better wear resistance, surface roughness, and lower noise level compared to uncoated tungsten carbide tools in the cutting of asbestos, WPC, LVL, and OSB. Among the coated carbide tools, the multilayer TiAlN/TiBON coated carbide tool is the highest in wear resistance and is proposed for cutting wood composites. The abrasives contained in the wood composite are important in the wearing of the tungsten carbide tools. The structures of the wood composites are important in determining the roughness of the machined surfaces. The noise level and roughness increase due to an increase in wear and should be a good indication for determining the wear of the coated and uncoated tungsten carbide tools [81]. This is probably related to the high hardness of the coating, low coefficient of friction, and the reported lubrication effect at high temperatures in cutting.

Ti-W-N/Ti-W and Cr-W-N/Cr-W multilayer coatings with different periods, as well as comparison of single-layer coatings Ti-W-N and Cr-W-N with a thickness from 0.5 to 2.0 μm, deposited on WC-Co substrates, were the subject of research conducted by Pinheiro et al. [41]. The results showed that both 1 μm thick single-layer coatings, tested when cutting OSB, increased wear resistance approximately 2.5 times compared to uncoated cutting tools. Using the same cutting parameters (as those used in industry), they showed that the best result was achieved by applying the Cr–W–N/Cr–W multilayer coatings (three layers). This coating increases the cutting ability of wood-based products by 500% compared to the uncoated cemented carbide tool.

Kong et al. [54] conducted a comparative study of the wear resistance of HSS and cemented carbide knives with a single-layer CrN and CrN/CrCN multi-layer coatings during rounding of wood. They showed that the service life of CrN/CrCN-coated tools increased by 170% (HSS) and 110% (cemented carbide) compared to uncoated tools and by approximately 33% and 7%, respectively, compared to CrN-coated tools. Two important conclusions can be drawn from this work: a harder substrate contributes to an increase in durability more than a lower hardness substrate when covered with the same coating, and a multi-layer coating gives a greater increase in durability compared to a single-layer coating. A similar result was obtained for the same multilayer coating on HSS knives [71].

Figure 2 shows the parts of two planer knives, without a coating (top) and with a CrN/CrN coating (bottom) [82]. The left side shows the knife wear after a distance of 77 km of machining pinewood at feed speed—90 m/min, cutting speed—36 m/s and cutting depth 2 mm. The right side shows a fragment of the knife after sharpening, without wood machining. The tests were performed using Weinig Hydromat 22A.

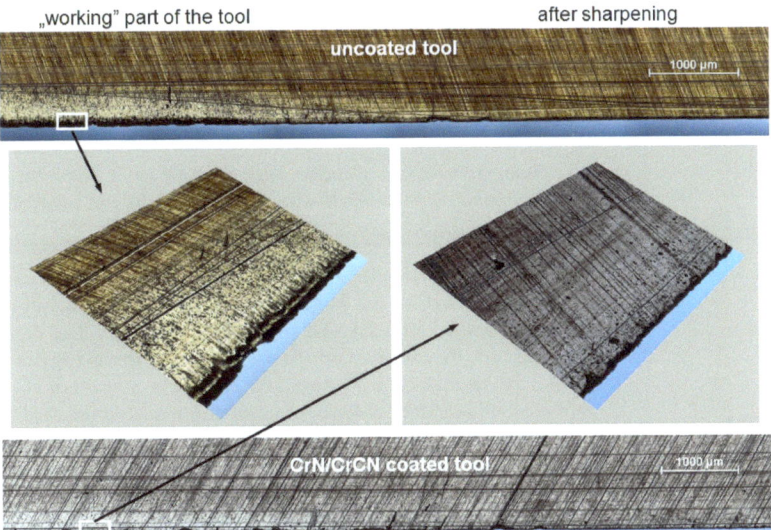

Figure 2. Fragments of uncoated HS6-5-2 steel knives (**top**) and CrN/CrCN (**bottom**). The left part after processing pine wood; the right part, after sharpening, did not process the wood [82].

Plastic deformation caused by high temperatures during cutting is visible on the edge of the uncoated tool. The blade was heavily worn after the machining test. On the edge of the CrN/CrCN-coated tool, slight wear is observed compared to the uncoated tool. In the middle part of Figure 2 is an enlargement of the edges of both knives. In the case of a tool with a CrN/CrCN coating, wear and delamination of the coating on the edge are hardly visible.

The test conditions favored an increase in temperature on the cutting edge and on the rake face. This leads to accelerated wear of the tools, resulting in, e.g., an increase in power consumption by the milling machine, and on the other hand, a deterioration in the quality of the processed material. Dependent on the type of processed wood, the wear rate and its symptoms may vary.

Tool wear is observed in the rounding of the edge of the blade and the abrasion of the rake face of the knife. Figure 3 shows a 3D view of the knife blade without a coating ready to work (Figure 3a), as well as the knife blade without coating (Figure 3b) and the knife blade with a CrN/CrCN multilayer coating (Figure 3c) after planing dry beech wood. The cutting path in this case was 6000 m. After sharpening, the uncoated knives had a blade radius of about 7 µm, while the coated knives had a slightly larger blade radius of about 10 µm. After cutting beech wood along a path of 6000 m, the radius of the knife blade without the coating increased more than three times and amounted to 24 µm. Under the same cutting conditions, the tip radius of the coated knife was approximately two times greater, at 20 µm.

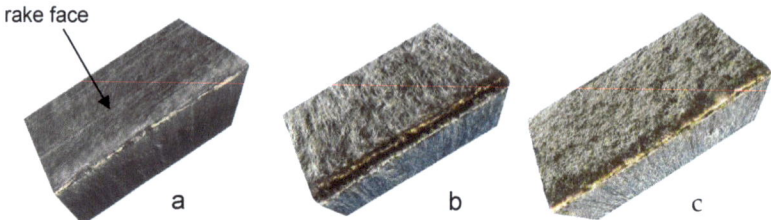

Figure 3. 3D image of cutting edge of planer knife without the coating after sharpening (**a**), the cutting edge of planer knife without the coating (**b**), and with CrCN/CrN multilayer coating (**c**) after 6000 m of dry beech wood cutting [83].

The rake face (Figure 3a) shows a different surface morphology in the blade working zone, caused by the temperature increased by machining, higher than the tempering temperature of the HS6-5-2 steel, i.e., above 550 °C (see Figure 2). This is represented by an image of the rake face profile perpendicular to the edge of the blade. There is visible wear of the knife rake face without the coating as a result of abrasion of the knife material. The coated knife does not show this type of wear (Figure 4). This indicates the good anti-wear properties of the coatings. The coatings also restrict the heat flow to the tool [50], which can reduce or eliminate the tempering effect of the tool material. High temperature causes plastic deformation of the material in the area of the blade edge, which promotes edge rounding and accelerates tool wear, and also deteriorates the quality of the machined wood surface. When sharpening the tool, restoring its original cutting properties, the material with mechanical properties changed by temperature should be removed. In the case of the tested knives made of HS6-5-2 steel, it is even 0.3 mm. This requires both an increase in the tool regeneration time and a reduction in the number of such processes, resulting in an increase in machining costs. In the case of knives with coatings, no negative influence of temperature and no reduction in the hardness of the tool material was observed.

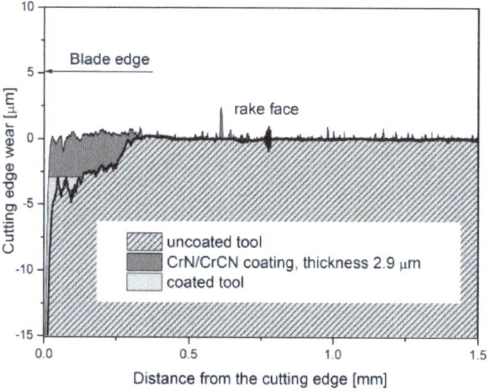

Figure 4. The cutting edge profile of planer knife with CrN/CrCN multilayer coating after 6000 m of dry beech wood cutting [83].

The greater efficiency of tools with a CrN/CrCN multilayer coating is manifested in the lower surface area of the blade wear in relation to the uncoated knife. For such selected tool-machined material systems and cutting process parameters, the surface area of the blade wear for the coated knife is approximately 25% lower, which translates into approximately three times longer cutting distance.

The analysis of the data presented in Tables A1–A3 shows that they are very diverse. The substrate, coatings, material treatments, and the material itself are different, as well as the presentation and evaluation of the results.

The records in the above Tables are arranged according to the sequential year of publication. It is not possible to unambiguously observe an increase in durability due to the variety of coatings, types of material being processed, and probably different machining parameters. One can also notice the important role of the deposition method and the tool material used as a substrate for the coating. The above data confirm that there is no universal coating that will be equally effective in the machining of various materials. The coating design process should take into account the type of material being processed, the machining parameters, and the type of tool to be used.

Nevertheless, based on the results presented in Tables A1–A3, it can be noted that:

- A comparison of various coatings operating in the same conditions shows the advantage of CrN both in the treatment of beech wood [53] and OSB [24,71];
- Worse performance properties of Cr_2N coatings compared to CrN [24,84] were confirmed, which may be caused by their greater brittleness and worse adhesion to the substrate;
- A comparison of the wear resistance of single-layer coatings was presented by Kazlauskas et al. [85]. Binary CrN and ternary AlCrN and TiCN coatings deposited on WC-Co tools (substrates) during the milling of oak wood showed improved wear resistance of cutters by factors of 3.0, 1.9, and 1.7, respectively. They also found that CrN was characterized by the best adhesion among the tested coatings and a low and stable coefficient of friction. They stated that low-friction coatings exhibited better wear resistance;
- Ti-based coatings have inferior wear resistance compared to Cr-based coatings. This applies to both magnetron- and arc-formed coatings. The Ti-W-N/Ti-W coating is characterized by twice the tool wear area compared to the Cr-W-N/Cr-W coating in the milling of OSB and particle board [41]. The wear of the tool covered by TiCN and CrN coating is smaller by 1.6 and 2.0, respectively [57,65]. The above conclusions are confirmed by Kazlauskas et al. [85];
- Multilayer coatings have better properties compared to single-layer coatings. CrN and CrN/CrCN coatings were deposited on a substrate made of M2 steel and cemented carbide and the timber boles were rounded [54]. Two significant effects were observed:

    (a) The multilayer coating was characterized by a greater wear resistance of about 170%, while in the case of CrN coatings the increase was about 100%. These results relate to the tests carried out on the M2 steel tool coating system. Cemented carbide tools with coatings were characterized by a slightly lower durability of approx. 110% (CrN/CrCN) and 100% (CrN). It should be noted, however, that uncoated tools made of M2 steel rounded 15 boles, and cemented carbide tools 188 boles;

    (b) The comparison of these coatings shows that in M2 steel tools with CrN/CrCN coatings the service time is about 33% higher compared to the CrN coating, and in the case of a cemented carbide tool the increase is about 7% [54]. This confirms the reports of many authors about the importance of the substrate;

- A comparison of the edge recession of tungsten carbide tools with TiAlN monolayer coating and TiAlN/TiSiN and TiAlN/TiBON multilayer coatings in milling different wood composites indicates a greater edge recession of the TiAlN and TiAlN/TiSiN coatings [81]. Among the tested coatings, the smallest edge recession was demonstrated by the TiAlN/TiBON multilayer coating, despite the fact that its hardness was the lowest among the tested coatings, 2700 HV. The TiAlN/TiSiN coating, with a hardness of 3600 HV and the highest coefficient of friction (0.9), showed greater edge recession than the TiAlN single-layer coating. The results of these tests confirm that a lower coefficient of friction is more important than hardness.

5.4. Substrates Used for Testing the Properties of the Coatings

There are basically two types of substrates in Table A1, 60SMD8 alloy steel with a hardness of 57–59 HRC [26,53], commonly used for cutting tools for peeling lathes, and 90MCV8 steel, hardness of 63–65 HRC, suitable for cutting and punching tools, etc., and cemented carbide with a hardness of about 17 GPa.

The substrate has a major influence on the performance of the coating. Chekour et al. considered the effect of the nitriding conditions of the 32CrMoV13 steel substrate as a tool for the peeling of beech [23]. It was found that in most cases the nitrided layer is about 200 nm thick and the surface hardness is about 1000 HV, twice that of unnitrided steel. Tool tests during peeling of the beech showed that the nitrided tool in a mixture with a small concentration of nitrogen performs like an unnitrided tool. Duplex-treated tools (nitriding and CrN coating) performed better than nitrided tools and allowed almost twice as much processing as untreated tools. Duplex processing (nitriding and CrN coating) of low-alloy steel, commonly used in the production of knives for the wood industry, allowed over 70% reduction of knives' edges [37].

5.5. Scientific Centers

One can indicate the scientific centers dealing with the production, implementation and testing of tools modified with coatings for processing of wood. Based on SCOPUS and Web of Science databases, one can indicate the dominant position of France [14,20–24,26,41,48,50–53,66–68,84,86] in this challenge but Germany [49,76], Japan [40,48,68,87], USA [71,87,88], UAE [28,29,71,87] and Poland [21,53,55–57,69–74,89–92] also deal with this subject extensively. Also, scientists from such countries as: Belarus [93–96], China [54,75], Finland [46], Indonesia [40,48,68,81], Italy [39], Lithuania [44,65,85,97–99], Portugal [41–100], Slovakia [94], and Switzerland [33] have significant achievements in research and dissemination of knowledge about modified protective coatings for woodworking tools.

6. Conclusions

The history of thin hard coatings goes back to the 1970s. The possibilities of new tool modification techniques were verified in laboratories and in industry, initially for metalworking, and somewhat later also for woodworking. Wood is a specific material; it is characterized by strong fluctuating local anisotropy, porosity, and a hygroscopic nature. Hence, the demands placed on woodworking tools are particularly high. Thanks to the progress in surface treatment, it is possible to obtain tools that meet the requirements of customers to a greater extent than before. Knowledge about the applied modifications of tools is becoming more and more available, as evidenced by the growing number of scientific publications, as well as the multinational composition of research teams. The main difficulty in analyzing the results is:

- Many articles are related to the processing of fiberboard or very dense homogeneous wood, the most homogeneous and isotropic materials. The processing of other materials, softwood and wood-based materials, does not give unequivocal results in the durability of the tools. It seems that the most important factors in modifying a wood knife are low friction value, good abrasion resistance, good coating adhesion, and thermal resistance. Most of the tested coatings belong to the simplest, two- or three-element systems. Only in some cases have more complex coatings, such as quaternary systems or multi-layer coatings with different structures, been investigated. The test results indicate that the latter have better wear resistance, but standardized tests should be performed to confirm this.

The analysis of the test results for the properties of coatings and the durability of woodworking tools covered with these coatings allows for the presentation of several conclusions:

- A comparison of various coatings operating in the same conditions shows the advantage of CrN both in the treatment of beech wood and OSB. This is probably due to the

lower brittleness of the CrN coatings, relatively low coefficient of friction, and very good adhesion to the substrate;
- Chromium-based coatings have better wear resistance compared to titanium-based coatings;
- Multilayer coatings are characterized by better wear resistance compared to single-layer coatings. This may be related to the reduction of crack propagation at the phase and grain boundaries. As in the case of a single-layer coating, higher wear resistance is found in coatings that exhibit a lower friction coefficient and are less brittle. The results of the tests confirm that a lower coefficient of friction is more important than hardness;
- The type of tool material used, and its possible thermochemical treatment, have a great influence on tool life. Increasing the hardness of the tool increases its productivity, although not always its durability.

Future Directions

The field of wood-cutting tools still requires further research. It seems that future research should focus on selecting one method of coating formation on a clearly defined substrate (tool). The choice of the tested multilayer coating (coatings) showing the best anti-wear properties is much more complex. It should be characterized by high hardness, good adhesion to the substrate, stability at elevated temperatures, relatively low coefficient of friction, and chemical inertness. Processing tests of the selected type of wood (wood-based) material should be limited to one specific species and the same series of processing parameters. It is also important to present the research results so that they can be compared with the references.

Author Contributions: Conceptualization, B.W. and A.G.; methodology, B.W. and A.G.; software, B.W.; validation, B.W.; formal analysis, A.G.; investigation, B.W. and A.G.; resources, B.W. and A.G.; data curation, B.W. and A.G.; writing—original draft preparation, B.W.; writing—review and editing, A.G.; visualization, B.W..; supervision, A.G.; project administration, A.G.; funding acquisition, A.G. All authors have read and agreed to the published version of the manuscript.

Funding: This research was part-financed by the NATIONAL CENTER FOR SCIENCE AND DEVELOPMENT of Poland within the 3rd strategic competition of the "Environment. Agriculture and Forestry" Program—BIOSTRATEG. No. BIOSTRATEG3/344303/14/NCBR/2018.

Institutional Review Board Statement: Not applicable.

Informed Consent Statement: Not applicable.

Data Availability Statement: Not applicable.

Conflicts of Interest: The authors declare no conflict of interest.

Abbreviations

MDF	medium density fiberboard
OSB	Oriented Strand Board
ta-C	tetragonal carbon
WPC	wood plastic composite
LVL	laminated veneer lumber
GRC	Glass-reinforced concrete

Appendix A

Table A1. Magnetron sputtering. Protective coatings applied to woodworking tools, synthetic information on the coatings used, the materials processed, the type of substrate and the effect of its modification.

Coating/Structure	Substrate	Type of Machining	Machining Material	Results	Year	Reference
W-C:H (DLC with WC precipitations) CrN	60SMD8	peeling	beech wood	reduction of the knives' edges by 38% (W-C:H) and 52% (CrN)	1999	[53]
CrN W–C:H(DLC)	60SMD8	peeling	beech wood	cutting edge reduction up to: DLC—60%, CrN—130%.	1999	[26]
CrN Cr_2N	Carbide	cutting	OSB	reduction of the knives' edges 52% (CrN) and 40% (Cr_2N)	2000	[84]
CrN Cr_2N	Carbide	cutting	OSB	service life four times higher (CrN), about 1.8 times higher (Cr_2N)	2001	[24]
CrN	32CrMoV13 Nitrided	peeling	beech wood	increase the service of the tool by a factor of 1.3	2003	[23]
CrN	Carbide	routing	OSB	decrease in nose width about 64%	2003	[51]
CrN	90MCV8	peeling	MDF	reduction in the wear of the edge by about 50%	2005	[14]
CrN	different carbidess	milling	OSB	dependent on type of substrates	2005	[52]
CrAlN	Carbide	routing	MDF	increase up to 2.5 times more than unmodified ones	2007	[66]
CrAlN	90CrMoV8	peeling	beech wood	reduction in the wear of the edge by about 50%	2009	[37]
AlCrN	WC-2% Co	routing	MDF, (M) standard, and (E) fireproof	max decrease in nose width about 25%—MDF(M), 40% MDF(E) and 44% MDF	2009	[83]
CrSiN	WC-2% Co	routing	MDF	max decrease in nose width about 33%	2009	[38]

Table A1. Cont.

Coating/Structure	Substrate	Type of Machining	Machining Material	Results	Year	Reference
Ti-W-N/Ti-W Cr-W-N/Cr-W	WC+4%Co	milling	OSB	reduction in average wear area: Ti based coatings—to 54%, Cr—to 100%,	2009	[41]
Ti-W-N/Ti-W Cr-W-N/Cr-W	WC+4%Co	milling	particle board	reduction in average wear area: Ti based coatings—to 215%, Cr—to 460%	2009	[41]
TiAlN TiAlN/aCN	Carbide	cutting	chipboard	max. increase by 23% (TiAlN/aCN)	2020	[89]
TiN/AlTiN TiAlN/a-C:N	Different carbides	cutting	chipboard	max. increase by 56% (TiN/TiAlN)	2021	[90]

Table A2. Cathodic arc evaporation. Protective coatings applied to woodworking tools, synthetic information on the coatings used, the materials processed, the type of substrate and the effect of its modification.

Coating/Structure	Substrate	Type of Machining	Machining Material	Results	Year	Reference
TiN (Ti,Zr)N	60SMD8 90WDCV	peeling	beech wood	cutting edge reduction up to: Ti based coatings 17%,	1999	[26]
CrN/CrCN	HS6-5-2	planing	pine wood	reduction in average wear area to 170%	2011	[71]
Cr_2N/CrN	HS6-5-2	cutting	pine wood	reduction in average wear area of 60%	2011	[69]
TiCN CrN DLC	K01–K20	milling	wood panel oaken scantlings glued by polyvinyl acetate	wear compared to the uncoated cutters. TiCN—smaller by 1.6 × DLC—smaller by 1.9 × CrN—smaller twice	2015	[65]
ZrN MoN	WC + Co	milling	particle board	reduction in volume wear to 150% (MoN) and 110% (ZrN)	2016	[93]
TiAlN	K10	cutting	mersawa wood fiberboard, particleboard, GRC	edge recession reduction by factor: 0.27 0.60 0.33 0.38	2016	[40]
TiAlN/TiSiN	K10	cutting	mersawa wood fiberboard particleboard GRC	edge recession reduction by factor: 0.38 0.78 0.43 0.54	2016	[40]

Table A2. Cont.

Coating/ Structure	Substrate	Type of Machining	Machining Material	Results	Year	Reference
TiAlN/TiBON	K10	cutting	mersawa wood fiberboard particleboard GRC	edge recession reduction by factor: 0.62 1.13 0.83 1.10	2016	[40]
CrN CrN/CrCN	M2 steel	cutting	timber	improvement in the tool durability of 170% (CrN/CrCN), 100% (CrN)	2018	[54]
CrN CrN/CrCN	cemented carbide	rounding	timber	improvement in the tool durability of 110% (CrN/CrCN), 100% (CrN)	2018	[54]
TiAlN	K10 tungsten carbide tool	milling	asbestos WPC LVL OSB	edge recession reduction by factor: 2.12 1.09 1.5 1.54	2019	[81]
TiAlN/TiSiN	K10 tungsten carbide tool	milling	asbestos WPC LVL OSB	edge recession reduction by factor: 1.08 0.84 1.35 1.33	2019	[81]
TiAlN/TiBON	K10 tungsten carbide tool	milling	asbestos WPC LVL OSB	edge recession reduction by factor: 2.33 3.18 4.0 3.67	2019	[81]
CrN/CrCN	HS6-5-2	planing	pine wood	improvement in the tool durability of 142%	2020	[91]
AlCrBN	HS6-5-2	planing	pine wood	improvement in the life service by 205%	2021	[92]

Table A3. Physical vapor deposition (PVD). Protective coatings applied to woodworking tools, synthetic information on the coatings used, the materials processed, the type of substrate and the effect of its modification.

Coating/ Structure	Substrate	Type of Machining	Machining Material	Results	Year	Reference
ta-C	cemented tungsten carbide	milling	melamine laminated particle board	2.5-fold lifetime increase	1999	[49]
TiN	SKH 51	sawing	oil palm afina sugi	tool wear 10% decrease 25% increase 64% increase	2006	[47]

Table A3. Cont.

Coating/Structure	Substrate	Type of Machining	Machining Material	Results	Year	Reference
TiN	P30	cutting	hardboard, cement board	life time increase 30–45%	2008	[48]
TiAlN/TiBN, TiAlN/TiSiN, TiAlN/CrAlN TiAlN	K10	milling	particle board	multilayer-coated tools experienced a smaller amount of delamination wear than the monolayer-coated tool. The best multilayer coating was TiAlN/CrAlN	2010	[68]
CrN, AlTiN, TiAlN, TiCN, and CrN	WC-Co	milling	oak wood	improvement in wear resistance by factors of: 3.0 (CrN), 1.9 (AlCrN), 1.7 (TiCN)	2022	[85]

References

1. Available online: www.Kadur.com (accessed on 31 August 2022).
2. Dembiński, C.; Potok, Z.; Kučerka, M.; Kminiak, R.; Očkajová, A.; Rogoziński, T. The Flow Resistance of the Filter Bags in the Dust Collector Operating in the Line of Wood-Based Furniture Panels Edge Banding. *Appl. Sci.* **2022**, *12*, 5580. [CrossRef]
3. Dong, W.; Xiong, X.; Ma, Y.; Yue, X. Woodworking Tool Wear Condition Monitoring during Milling Based on Power Signals and a Particle Swarm Optimization-Back Propagation Neural Network. *Appl. Sci.* **2021**, *11*, 9026. [CrossRef]
4. Mračková, E.; Schmidtová, J.; Marková, I.; Jaďuďová, J.; Tureková, I.; Hitka, M. Fire Parameters of Spruce (*Picea abies Karst.* (L.)) Dust Layer from Different Wood Technologies Slovak Case Study. *Appl. Sci.* **2022**, *12*, 548. [CrossRef]
5. Orlowski, K.A.; Dudek, P.; Chuchala, D.; Blacharski, W.; Przybyliński, T. The Design Development of the Sliding Table Saw Towards Improving Its Dynamic Properties. *Appl. Sci.* **2020**, *10*, 7386. [CrossRef]
6. Pędzik, M.; Rogoziński, T.; Majka, J.; Stuper-Szablewska, K.; Antov, P.; Kristak, L.; Kminiak, R.; Kučerka, M. Fine Dust Creation during Hardwood Machine Sanding. *Appl. Sci.* **2021**, *11*, 6602. [CrossRef]
7. Sporek, M.; Sporek, K.; Stebila, J.; Kučerka, M.; Kminiak, R.; Lubis, M.A.R. Assessment of the Mass and Surface Area of the Scots Pine (*Pinus sylvestris* L.) Needles. *Appl. Sci.* **2022**, *12*, 8204. [CrossRef]
8. Sydor, M.; Mirski, R.; Stuper-Szablewska, K.; Rogoziński, T. Efficiency of Machine Sanding of Wood. *Appl. Sci.* **2021**, *11*, 2860. [CrossRef]
9. Sydor, M.; Pinkowski, G.; Kučerka, M.; Kminiak, R.; Antov, P.; Rogoziński, T. Indentation Hardness and Elastic Recovery of Some Hardwood Species. *Appl. Sci.* **2022**, *12*, 5049. [CrossRef]
10. Norrby, N.; Johansson, M.P.; M'Saoubi, R.; Odén, M. Pressure and temperature effects on the decomposition of arc evaporated $Ti_{0.6}Al_{0.4}N$ coatings in continuous turning. *Surf. Coat. Technol.* **2012**, *209*, 203–207. [CrossRef]
11. Grobelny, T. Thermo-Mechanical Condition of the Tool Blade in the Process of Milling Wood and Wood Materials. Ph.D. Thesis, SGGW, Warsaw, Poland, 1999.
12. Lopez De Lacalle, L.N.; Lamikiz, A.; De Larrinova, J.F.; Azkona, I. Advanced Cutting Tools. In *Machining of Hard Materials*; Davim, J.P., Ed.; Springer-Verlag: London, UK, 2011; pp. 33–85.
13. Yates, J.M. Is there life after tungsten carbide? *Wood Wood Prod.* **1987**, *92*, 77–87.
14. Boyle, G.R. PCD replacing carbide in woodworking applications. *Cut. Tool Eng.* **1983**, *35*, 60–61.
15. Aytacoglu, M.E. PCD cutting tools gaining momentum. *Cut. Tool Eng.* **1983**, *35*, 49–50.
16. Heimbrand, E. Machining wood products with PCD. *Ind. Diam. Rev.* **1985**, *45*, 187–190.
17. Herbert, S. Wood products latch on to PCD. *Ind. Diam. Rev.* **1984**, *44*, 159–162.
18. Klamecki, B.E. A review of wood cutting tool wear literature. *Holz Als Roh-Und Werkst.* **1979**, *37*, 265–276. [CrossRef]
19. Thibaut, B.; Denaud, L.; Collet, R.; Marchal, R.; Beauchene, J.; Mothe, F.; Méausoone, P.J.; Martin, P.; Larricq, P.; Eyma, F. Wood machining with a focus on French research in the last 50 years. *Ann. For. Sci.* **2016**, *73*, 163–184. [CrossRef]
20. Rudnicki, J.; Beer, P.; Sokołowska, A.; Marchal, R. Low-temperature ion nitriding used for improving the durability of the steel knives in the wood rotary peeling. *Surf. Coat. Technol.* **1998**, *107*, 20–23. [CrossRef]
21. Nouveau, C.; Steyer, P.; Mohan Rao, K.R.; Lagadrillere, D. Plasma nitriding of 90CrMoV8 tool steel for the enhancement of hardness and corrosion resistance. *Surf. Coat. Technol.* **2011**, *205*, 4514–4520. [CrossRef]
22. Chekour, L.; Nouveau, C.; Chala, A.; Djouadi, M.A. Duplex treatment of 32CrMoV13 steel by ionic nitriding and triode sputtering: Application to wood machining. *Wear* **2003**, *255*, 1438–1443. [CrossRef]

23. Nouveau, C.; Djouadi, M.A.; Decès-Petit, C.; Beer, P.; Lambertin, M. Influence of Cr_xN_y coatings deposited by magnetron sputtering on tool service life in wood processing. *Surf. Coat. Technol.* **2001**, *142–144*, 94–101. [CrossRef]
24. Labidi, C.; Collet, R.; Nouveau, C.; Beer, P.; Nicosia, S.; Djouadi, M.A. Surface treatments of tools used in industrial wood machining. *Surf. Coat. Technol.* **2005**, *200*, 118–122. [CrossRef]
25. Anon, J.C.R. Modern cutting tools adapt to changing woodworking needs. *Wood Wood Prod.* **1984**, *89*, 201–203.
26. Djouadi, M.A.; Beer, P.; Marchal, R.; Sokolowska, A.; Lambertin, M.; Precht, W.; Nouveau, C. Antiabrasive coatings: Application for wood processing. *Surf. Coat. Technol.* **1999**, *116–119*, 508–516. [CrossRef]
27. Naylor, A.; Hackney, P. A Review of Wood Machining Literature with a Special Focus on Sawing. *BioResources* **2013**, *8*, 3122–3135. [CrossRef]
28. Sheikh-Ahmad, J.Y. Effect of Cutting Edge Geometry on Thermal Stresses and Failure of Diamond Coated Tools. *Procedia Manuf.* **2015**, *1*, 663–674. [CrossRef]
29. Sheikh-Ahmad, J.Y.; Morita, T. Tool coatings for wood machining: Problems and prospects. *For. Prod. J.* **2002**, *52*, 43–51.
30. Gogolewski, P.; Klimke, J.; Krell, A.; Beer, P. Al_2O_3 tools towards effective machining of wood-based materials. *J. Mater. Process. Technol.* **2009**, *209*, 2231–2236. [CrossRef]
31. Guo, X.L.; Cao, P.X.; Liu, H.N.; Teng, Y.; Guo, Y.; Wang, H. Tribological Properties of Ceramics Tool Materials in Contact with Wood-Based Materials. *Adv. Mat. Res.* **2013**, *764*, 65–69. [CrossRef]
32. Zhu, Z.; Guo, X.; Na, B.; Liang, X.; Ekevad, M.; Ji, F. Research on cutting performance of ceramic cutting tools in milling high density fiberboard. *Wood Res.* **2017**, *62*, 125–138.
33. Eblagon, F.; Ehrle, B.; Graule, T.; Kuebler, J. Development of silicon nitride/silicon carbide composites for wood-cutting tools. *J. Eur. Ceram. Soc.* **2007**, *27*, 419–428. [CrossRef]
34. Ugulino, B.; Hernandez, R.E. Assessment of surface properties and solvent-borne coating performance of red oak wood produced by peripheral planing. *Eur. J. Woo Prod.* **2017**, *75*, 581–593. [CrossRef]
35. Blugan, G.; Strehler, V.; Vetterli, M.; Ehrle, B.; Duttlinger, R.; Blösch, P.; Kuebler, J. Performance of lightweight coated oxide ceramic composites for industrial high speed wood cutting tools: A step closer to market. *Ceram. Int.* **2017**, *43*, 8735–8742. [CrossRef]
36. Beer, P.; Gogolewski, P.; Klimke, J.; Krell, A. Tribological Behaviour of Sub-micron Cutting-ceramics in Contact with Wood-based Materials. *Tribol. Lett.* **2007**, *27*, 155–158. [CrossRef]
37. Beer, P.; Rudnicki, J.; Ciupinski, L.; Djouadi, M.A.; Nouveau, C. Modification by composite coatings of knives made of low alloy steel for wood machining purposes. *Surf. Coat. Technol.* **2003**, *174–175*, 434–439. [CrossRef]
38. Benlatreche, Y.; Nouveau, C.; Aknouche, H.; Imhoff, L.; Martin, N.; Gavoille, J.; Rousselot, C.; Rauch, J.Y.; Pilloud, D. Physical and Mechanical Properties of CrAlN and CrSiN Ternary Systems for Wood Machining Applications. *Plasma Process. Polym.* **2009**, *6*, S113–S117. [CrossRef]
39. Faga, M.G.; Settineri, L. Innovative anti-wear coatings on cutting tools for wood machining. *Surf. Coat. Technol.* **2006**, *201*, 3002–3007. [CrossRef]
40. Fahrussiam, F.; Praja, I.A.; Darmawan, W.; Wahyudi, I.; Nandika, D.; Usuki, H.; Koseki, S. Wear characteristics of multilayer-coated cutting tools in milling wood and wood-based composites. *Tribol. Ind.* **2016**, *38*, 66–73.
41. Pinheiro, D.; Vieira, M.T.; Djouadi, M.A. Advantages of depositing multilayer coatings for cutting wood-based products. *Surf. Coat. Technol.* **2009**, *203*, 3197–3205. [CrossRef]
42. Vahlas, C. Chemical vapor deposition of metals: From unary systems to complex metallic alloys. In *Surface Properties and Engineering of Complex Metallic Alloys*; Belin-Ferre, E., Ed.; World Scientific Publishing Company Co., Pte. Ltd.: Singapore, 2010; Volume 3, pp. 49–82.
43. Gulbiński, W. Physical vapor deposition of thin film coatings. In *Surface Properties and Engineering of Complex Metallic Alloys*; Belin-Ferre, E., Ed.; World Scientific Publishing Company Co., Pte. Ltd.: Singapore, 2010; Volume 3, pp. 83–92.
44. Bendikiene, R.; Keturakis, G.; Pilkaite, T.; Pupelis, E. Wear Behaviour and Cutting Performance of Surfaced Inserts for Wood Machining. *Stroj.Vestn.-J. Mech. E.* **2015**, *61*, 459–464. [CrossRef]
45. Valleti, K.; Rejin, C.; Joshi, S.V. Factors influencing properties of CrN thin films grown by cylindrical cathodic arc physical vapor deposition on HSS substrates. *Mater. Sci. Eng. A* **2012**, *545*, 155–161. [CrossRef]
46. Osenius, S.; Korhonen, A.S.; Sulonen, M.S. Performance of TiN-coated tools in wood cutting. *Surf. Coat. Technol.* **1987**, *33*, 141–151. [CrossRef]
47. Okai, R.; Tanaka, C.; Iwasaki, Y. Influence of mechanical properties and mineral salts in wood species on tool wear of high-speed steels and stellite-tipped tools—Consideration of tool wear of the newly developed tip-inserted band saw. *Holz Als Roh-Und Werkst.* **2006**, *64*, 45–52. [CrossRef]
48. Darmawan, W.; Usuki, H.; Quesada, J.; Marchal, R. Clearance wear and normal force of TiN-coated P30 in cutting hardboards and wood-chip cement boards. *Holz Als Roh-Und Werkst.* **2008**, *66*, 89–97. [CrossRef]
49. Endler, I.; Bartsch, K.; Leonhardt, A.; Scheibe, H.J.; Ziegele, H.; Fuchs, I.; Raatz, C. Preparation and wear behaviour of woodworking tools coated with superhard layers. *Diam. Relat. Mater.* **1999**, *8*, 834–839. [CrossRef]
50. Kusiak, A.; Battaglia, J.L.; Marchal, R. Influence of CrN coating in wood machining from heat flux estimation in the tool. *Int. J. Thermal Sci.* **2005**, *44*, 289–301. [CrossRef]
51. Nouveau, C.; Djouadi, M.A.; Decès-Petit, C. The influence of deposition parameters on the wear resistance of Cr_xN_y magnetron sputtering coatings in routing of oriented strand board. *Surf. Coat. Technol.* **2003**, *174–175*, 455–460. [CrossRef]

52. Nouveau, C.; Jorand, E.; Decès-Petit, C.; Labidi, C.; Djouadi, M.A. Influence of carbide substrates on tribological properties of chromium nitride coatings: Application to wood machining. *Wear* **2005**, *258*, 157–165. [CrossRef]
53. Beer, P.; Djouadi, M.A.; Marchal, R.; Sokolowska, A.; Lambertin, M.; Czyzniewski, A.; Precht, W. Antiabrasive coatings in a new application—Wood rotary peeling process. *Vacuum* **1999**, *53*, 363–366. [CrossRef]
54. Kong, Y.; Tian, X.; Gong, C. Enhancement of toughness and wear resistance by CrN/CrCN multilayered coatings for wood processing. *Surf. Coat. Technol.* **2018**, *344*, 204–213. [CrossRef]
55. Niedzielski, P.; Miklaszewski, S.; Beer, P.; Sokolowska, A. Tribological properties of NCD coated cemented carbides in contact with wood. *Diam. Rel. Mater.* **2001**, *10*, 1–6. [CrossRef]
56. Pancielejko, M.; Czyzniewski, A.; Gilewicz, A.; Zavaleyev, V.; Szymański, W. The cutting properties and wear of the knives with DLC and W-DLC coatings. deposited by PVD methods. applied for wood and wood-based materials machining. *Arch. Mater. Sci. Eng.* **2012**, *58*, 235–244.
57. Kaczorowski, W.; Batory, D.; Szamanski, W.; Niedzielski, P. Carbon-based layers for mechanical machining of wood-based materials. *Wood Sci. Technol.* **2012**, *46*, 1085–1096. [CrossRef]
58. Olaya, J.J.; Wei, G.; Rodil, S.E.; Muhl, S.; Bhushan, B. Influence of the ion–atom flux ratio on the mechanical properties of chromium nitride thin films. *Vacuum* **2007**, *81*, 610–618. [CrossRef]
59. Odén, M.; Ericsson, C.; Håkansson, G.; Ljungcrantz, H. Microstructure and mechanical behavior of arc-evaporated Cr–N coatings. *Surf. Coat. Technol.* **1999**, *114*, 39–51. [CrossRef]
60. Oden, M.; Almer, J.; Hakansson, G.; Olsson, M. Microstructure property relationships in arc-evaporated CrN coatings. *Thin Solid Films* **2000**, *377–378*, 407–412. [CrossRef]
61. Navinsek, B.; Panjan, P.; Milosev, I. Industrial applications of CrN (PVD) coatings. deposited at high and low temperatures. *Surf. Coat. Technol.* **1997**, *97*, 182–191. [CrossRef]
62. Rebholz, C.; Ziegele, H.; Leyland, A.; Matthew, A. Structure, mechanical and tribological properties of nitrogen-containing chromium coatings prepared by reactive magnetron sputtering. *Surf. Coat. Technol.* **1999**, *115*, 222–229. [CrossRef]
63. Wiklund, U.; Gunnars, J.; Hogmark, S. Influence of residual stresses on fracture and delamination of thin hard coatings. *Wear* **1999**, *232*, 262–269. [CrossRef]
64. Castanho, J.M.; Vieira, M.T. Effect of ductile layers in mechanical behaviour of TiAlN thin coatings. *J. Mater. Proc. Technol.* **2003**, *143–144*, 352–357. [CrossRef]
65. Kazlauskas, D.; Keturakis, G. Wear of TiCN, CrN and DLC coated tungsten carbide router cutters during oak wood milling. In Proceedings of the 20th International Scientific Conference: Mechanika 2015, Kaunas University of Technology, Kaunas, Lithuania, 23–24 April 2015.
66. Nouveau, C.; Labidi, C.; Ferreira Martin, J.P.; Collet, R.; Djouadi, M.A. Application of CrAlN coatings on carbide substrates in routing of MDF. *Wear* **2007**, *263*, 1291–1299. [CrossRef]
67. Nouveau, C.; Labidi, C.; Collet, R.; Benlatreche, Y.; Djouadi, M.A. Effect of surface finishing such as sand-blasting and CrAlN hard coatings on the cutting edge's pelling tools' wear resistance. *Wear* **2009**, *267*, 1062–1067. [CrossRef]
68. Darmawan, W.; Usuki, H.; Rahayu, I.S.; Gottlöber, C.; Marchal, R. Wear Characteristics of Multilayer-Coated Cutting Tools when Milling Particleboard. *For. Prod. J.* **2010**, *60*, 615–621. [CrossRef]
69. Warcholinski, B.; Gilewicz, A.; Ratajski, J. Cr$_2$N/CrN multilayer coatings for wood machining tools. *Tribol. Int.* **2011**, *244*, 1076–1082. [CrossRef]
70. Warcholinski, B.; Gilewicz, A.; Kuklinski, Z.; Myslinski, P. Hard CrCN/CrN multilayer coatings for tribological applications. *Surf. Coat. Technol.* **2010**, *204*, 2289–2293. [CrossRef]
71. Gilewicz, A.; Warcholinski, B.; Myslinski, P.; Szymanski, W. Anti-wear multilayer coatings based on chromium nitride for wood machining tools. *Wear* **2010**, *210*, 32–38. [CrossRef]
72. Warcholinski, B.; Gilewicz, A. Multilayer coatings on tools for woodworking. *Wear* **2011**, *271*, 2812–2820. [CrossRef]
73. Gilewicz, A.; Warcholinski, B.; Szymanski, W.; Grimm, W. CrCN/CrN + ta-C multilayer coating for applications in wood processing. *Tribol. Int.* **2013**, *57*, 1–7. [CrossRef]
74. Kot, M.; Rakowski, W.A.; Major, Ł.; Major, R.; Morgiel, J. Effect of bilayer period on properties of Cr/CrN multilayer coatings produced by laser ablation. *Surf. Coat. Technol.* **2008**, *202*, 3501–3506. [CrossRef]
75. Chang, Y.Y.; Yang, S.J.; Wang, D.Y. Structural and mechanical properties of TiAlN/CrN coatings synthesized by a cathodic-arc deposition process. *Surf. Coat. Technol.* **2006**, *201*, 4209–4214. [CrossRef]
76. Salje, E.W.; Stuehmeier, W. Milling particleboard with high hard cutting materials. In Proceedings of the Ninth International Wood Machining Seminar, Richmond, VA, USA, 10–12 October 1988; pp. 211–228.
77. Fuch, I.; Raatz, C. Study of wear behavior of specially coated (CVD.PACVD) cemented carbide tools while milling of wood-based materials. In Proceedings of the 13th International Wood Machining Seminar, Vancouver, BC, Canada, 17–19 June 1997; pp. 709–715.
78. Sheikh-Ahmad, J.Y.; Stewart, J.S. Performance of different PVD coated tungsten carbide tools in the continuous machining of particleboard. In Proceedings of the 12th International Wood Machining Seminar, Kyoto, Japan, 2–4 October 1995; pp. 282–291.
79. Li, Y.; Liu, Y.H.; Tian, Y.B.; Wang, Y.; Wang, J.L. Application of improved fireworks algorithm in grinding surface roughness online monitoring. *J. Manuf. Process.* **2022**, *74*, 400–412. [CrossRef]

80. Li, Y.; Liu, Y.H.; Wang, J.L.; Wang, Y.; Tian, Y.B. Real-time monitoring of silica ceramic composites grinding surface roughness based on signal spectrum analysis. *Ceram. Int.* **2022**, *48*, 7204–7217. [CrossRef]
81. Pangestu, K.P.T.; Darmawan, W.; Nandika, D.; Usuki, H. Cutting performance of multilayer coated tungsten carbide in milling of wood composites. *Int. Wood Prod. J.* **2019**, *10*, 78–85. [CrossRef]
82. Warcholinski, B.; Gilewicz, A. Coating for better endurance. In *FMD ASIA Solid Wood Panel Technology*; 2012; pp. 26–29.
83. Warcholiński, B.; Gilewicz, A.; Szymański, W.; Pinkowski, G. Improvment of durability of engineering tools for wood. *Surf. Eng.* **2011**, *2*, 73–80. (In Polish)
84. Djouadi, M.A.; Nouveau, C.; Beer, P.; Lambertin, M. Cr_xN_y hard coatings deposited with PVD method on tools for wood machining. *Surf. Coat. Technol.* **2000**, *133–134*, 478–483. [CrossRef]
85. Kazlauskas, D.; Jankauskas, V.; Kreivaitis, R.; Tuckute, S. Wear behaviour of PVD coating strengthened WC-Co cutters during milling of oak-wood. *Wear* **2022**, *498–499*, 204336. [CrossRef]
86. Benlatreche, Y.; Nouveau, C.; Marchal, R.; Ferreira Martins, J.P.; Aknouche, H. Applications of CrAlN ternary system in wood machining of medium density fibreboard (MDF). *Wear* **2009**, *267*, 1056–1061. [CrossRef]
87. Kato, C.; Bailey, J.A. Wear Characteristics of a Woodworking Knife with a Vanadium Carbide Coating only on the Clearance Surface (Back Surface). *Key Eng. Mater.* **1998**, *138–140*, 479–520. [CrossRef]
88. Makala, R.S.; Yoganand, S.N.; Jagannadham, K.; Lemaster, R.L.; Bailey, J. Diamond Coated WC Tools for Machining Wood and Particle Board. *MRS Online Proc. Libr.* **2001**, *697*, 819. [CrossRef]
89. Czarniak, P.; Szymanowski, K.; Kucharska, B.; Krawczyńska, A.; Sobiecki, J.R.; Kubacki, J.; Panjan, P. Modification of tools for wood based materials machining with TiAlN/a-CN coating. *J. Mater. Sci. Eng. B* **2020**, *257*, 114540. [CrossRef]
90. Kucharska, B.; Sobiecki, J.R.; Czarniak, P.; Szymanowski, K.; Cymerman, K.; Moszczyńska, D.; Panjan, P. Influence of Different Types of Cemented Carbide Blades and Coating Thickness on Structure and Properties of TiN/AlTiN and TiAlN/a-C:N Coatings Deposited by PVD Techniques for Machining of Wood-Based Materials. *Materials* **2021**, *14*, 2740. [CrossRef]
91. Nadolny, K.; Kapłonek, W.; Sutowska, M.; Sutowski, P.; Myśliński, P.; Gilewicz, A. Experimental Studies on Durability of PVD-Based CrCN/CrN-Coated Cutting Blade of Planer Knives Used in the Pine Wood Planing Process. *Materials* **2020**, *13*, 2398. [CrossRef] [PubMed]
92. Nadolny, K.; Kapłonek, W.; Sutowska, M.; Sutowski, P.; Myśliński, P.; Gilewicz, A.; Warcholiński, B. Moving towards sustainable manufacturing by extending the tool life of the pine wood planing process using the AlCrBN coating. *SM&T* **2021**, *28*, e00259.
93. Chayeuski, V.; Zhylinski, V.; Grishkevich, A.; Rudak, P.; Barcik, S. Influence of high energy treatment on wear of edges knives of wood-cutting tool. *MM Sci. J.* **2016**, *6*, 1519–1523. [CrossRef]
94. Latushkina, S.D.; Rudak, P.V.; Kuis, D.V.; Rudak, O.G.; Posylkina, O.I.; Piskunova, O.Y.; Kováč, J.; Krilek, J.; Barcík, S. Protective Woodcutting Tool Coatings. *Acta Univ. Agric. Silvic. Mendel. Brun.* **2016**, *64*, 835–839. [CrossRef]
95. Chayeuski, V.; Zhylinski, V.; Cernashejus, O.; Visniakov, N.; Mikalauskas, G. Structural and Mechanical Properties of the ZrC/Ni-Nanodiamond Coating Synthesized by the PVD and Electroplating Processes for the Cutting Knifes. *J. Mater. Eng. Perform.* **2019**, *28*, 1278–1285. [CrossRef]
96. Kuleshov, A.K.; Uglov, V.V.; Rusalsky, D.P.; Grishkevich, A.A.; Chayeuski, V.V.; Haranin, V.N. Effect of ZrN and Mo-N coatings and sulfacyyanization on wear of wood-cutting knives. *J. Frict. Wear.* **2014**, *35*, 201–209. [CrossRef]
97. Kazlauskas, D.; Jankauskas, V.; Tuckute, S. Research on tribological characteristics of hard metal WC-Co tools with TiAlN and CrN PVD coatings for processing solid oak wood. *Coating* **2020**, *10*, 632. [CrossRef]
98. Kazlauskas, D.; Jankauskas, V. Woodworking tools: Tribological problems and directions of solutions. In Proceedings of the 9th International Conference BALTTRIB'2017, Kaunas, Lithuania, 16–17 November 2017.
99. Kazlauskas, D.; Keturakis, G.; Jankauskas, V.; Andriušis, A. Investigation of TiCrN-Coated High Speed Steel Tools Wear during Medium Density Fiberboard Milling. *J. Frict. Wear.* **2021**, *42*, 124–129. [CrossRef]
100. Castanho, J.M.; Pinheiro, D.; Vieira, M.T. New Multilayer Coatings for Secondary Wood Products Cutting. *Mater. Sci. Forum.* **2004**, *455–456*, 619–622. [CrossRef]

Article

Indentation Hardness and Elastic Recovery of Some Hardwood Species

Maciej Sydor [1,*], Grzegorz Pinkowski [1], Martin Kučerka [2,*], Richard Kminiak [3], Petar Antov [4] and Tomasz Rogoziński [5]

1. Department of Woodworking Machines and Fundamentals of Machine Design, Faculty of Forestry and Wood Technology, Poznań University of Life Sciences, 60-637 Poznań, Poland; grzegorz.pinkowski@up.poznan.pl
2. Faculty of Natural Sciences, Matej Bel University, 974 09 Banská Bystrica, Slovakia
3. Department of Woodworking, Faculty of Wood Science and Technology, Technical University in Zvolen, 960 01 Zvolen, Slovakia; richard.kminiak@tuzvo.sk
4. Faculty of Forest Industry, University of Forestry, 1797 Sofia, Bulgaria; p.antov@ltu.bg
5. Department of Furniture Design, Faculty of Forestry and Wood Technology, Poznań University of Life Sciences, 60-637 Poznań, Poland; tomasz.rogozinski@up.poznan.pl
* Correspondence: maciej.sydor@up.poznan.pl (M.S.); martin.kucerka@umb.sk (M.K.)

Citation: Sydor, M.; Pinkowski, G.; Kučerka, M.; Kminiak, R.; Antov, P.; Rogoziński, T. Indentation Hardness and Elastic Recovery of Some Hardwood Species. *Appl. Sci.* **2022**, *12*, 5049. https://doi.org/10.3390/app12105049

Academic Editor: Giuseppe Lazzara

Received: 21 April 2022
Accepted: 15 May 2022
Published: 17 May 2022

Publisher's Note: MDPI stays neutral with regard to jurisdictional claims in published maps and institutional affiliations.

Copyright: © 2022 by the authors. Licensee MDPI, Basel, Switzerland. This article is an open access article distributed under the terms and conditions of the Creative Commons Attribution (CC BY) license (https://creativecommons.org/licenses/by/4.0/).

Abstract: The purpose of the study was to measure the Brinell hardness (HB) of six wood species and evaluate the ability to recover the depth of the imprint (self-re-deformation). Straight-grain clear samples of ash, beech, alder, birch, iroko, and linden wood were prepared. Measurements were made in the three main reference timber cross-sections: radial (R), tangential (T), and axial/longitudinal (L) and with two measuring loads of 30 kG and 100 kG (294.2 N and 980.7 N). The tested wood species could be classified into hard (ash, beech), medium-hard (alder, birch, iroko), and soft (linden) wood species. The HBs of each tested wood species differed in the cross-sections, i.e., side hardness (R, T) and end hardness (L). Higher HB values were obtained at 100 kG load in all species and all three cross-sections. The lowest influence of the measurement force value on the HB value was revealed for the soft wood species (linden: 107–118%). This influence was visible for the other five medium-hard and hard wood species, ranging from 125% to 176%. The percentage of temporary imprint in total imprint depth (x/H) varied from 12 to 33% (linden 12–18%—the lowest self-re-deformation ability; beech 25–33%—the highest self-re-deformation ability). The results of this study underline that the higher the density of the wood, the higher the Brinell hardness, and, simultaneously, the greater the measurement force used, the higher the Brinell hardness measured. The ability of self-re-deformation in wood's R and T cross-sections depends on the wood density and the measuring force used. In contrast, this ability only depends on the wood density in the L cross-section. Those observations imply that the compaction of the cell structure during side compression is mainly non-destructive, while the longitudinal deformation of the cell structure (the buckling of cell walls and fracture of ends of the cells) is to a great degree destructive and irreversible. These results can be used in the construction and furniture sectors, especially when designing products and planning the woodworking of highly loaded wood floors and furniture elements.

Keywords: wood hardness; Brinell hardness; indentation depth; plastic deformation; elastic deformation; imprint recovery; indentation recovery; alder; linden; birch; ash; iroko; beech

1. Introduction

Hardness is the ability of a material to resist localized deformation. Hardness test force can be applied by scratching, cutting, mechanical wear, bending, dynamic or static indentation. In static indentation hardness measurement methods, a non-deformable ball, pyramid, cone, cylinder, or needle-shaped indenters are applied [1–3]. The measured indentation projection area, total indentation area, indentation depth, or force needed to indent an indenter to the required depth is used for hardness calculation. The principal wood macro-indentation hardness testing approaches are Brinell, Vickers, Rockwell, Meyer, Knoop, Shore, Leeb, and Janka. The most widely used test procedures for measuring the hardness of wood materials are the Brinell–Mörath, Janka, and Monnin methods [4], which means that rounded intenders are preferred, those that do not crack the wood. The Brinell method for wood materials is standardized in EN 1534: 2020 [5], which describes the method of testing the hardness of wood floors using a 10 mm ball indenter, and a force of 1 kN reached in 15 ± 3 s, maintained for 25 ± 5 s and then released. The imprint size measurement is performed after the load is removed, so this method concerns only the plastic imprint size. The scientific literature presents a proposal to modify the Brinell method in measuring the hardness of wood. This proposal postulates calculating the wood hardness based on the depth of the imprint under load, including both elastic and plastic components of the imprint [6,7].

In our previous study [8], the hardness of wood materials was measured using the Brinell method, and, in addition to the size of the permanent imprint (which is a measure of Brinell hardness), we also analyzed an elastic component of the imprints, i.e., one that spontaneously disappears after removing the measuring force. In this way, we tested a feature of the wood in addition to hardness: the ability to self-shallow the imprint after removing the measuring force (self-re-deformation). Analyzing the hardness and self-re-deformation ability, we concluded that tested materials could be divided into soft (beech, pine, iroko), medium-hard (merbau, common oak, maple, red oak), and hard materials, e.g., high-density fiberboards (HDF), plywood. The highest relative value of the plastic imprint in total deformation, ranging from 79 to 83%, was observed in the soft materials tested. Values ranged from 72 to 76% in medium-hard materials and only about 65% in hard materials. Therefore, hard materials exhibited the highest ability, among the materials tested, to reduce the depth of deformation immediately after force removal. A measuring force of 30 kG was used in these tests, and the hardness was measured in one wood cross-section (side hardness). Wood, a material with cylindrical orthotropy, has three perpendiculars one to another, reference main cross-sections. The main cross-sections are related to the wood grain direction: the longitudinal (L) cross-section (also called axial), the radial (R), that of its secondary growth, and the tangential (T), orthogonal to both [9]. The so-called "end hardness", measured in the L cross-section, is higher than the "side hardness", measured in the cross-sections R and T, which are close to each other [9–11]. This study aimed to measure the hardness of the six hardwood species in all three main cross-sections and evaluate their tendency to self-re-deformation, that is, the self-executing flattening of the measuring ball imprint.

2. Materials and Methods

The research was carried out on six hardwoods. The test samples, 30 mm × 30 mm × 20 mm in size, were made of: alder, linden, birch, ash, iroko, and beech. All logs used to prepare the test samples had a regular structure (not eccentric). The test samples were clear and made from logs without any structural defects of the wood. The logs were primarily cut into lumber according to three principal anatomical planes of reference in the stem: radial (R), tangential (T), and axial/longitudinal (L) cross-sections (Figure 1). Twelve samples were made for each of the wood species tested.

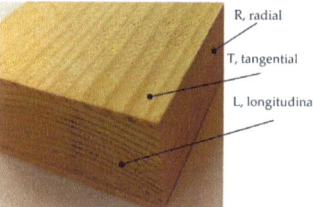

Figure 1. The clear test specimen oriented according to the main cross-sections in the stem.

After the primary cut, the planks were dried at a temperature lower than 60 °C. The test samples were cut and then conditioned at a temperature of 20 ± 2 °C and relative humidity of 65 ± 3% for three months. The moisture content of the test samples immediately before the hardness measurements was 10 ± 0.5%, and their average densities are given in Table 1.

Table 1. Wood species names and average densities of test samples used.

Species	Average Density (g/cm^3)
Alder (*Alnus glutinosa* (L.) Gaertn)	0.500
Linden (*Tilia europaea* L)	0.505
Birch (*Betula alba* L.)	0.595
Ash (*Fraxinus excelsior* L.)	0.660
Iroko (*Milicia excelsa* (Welw.) CC Berg)	0.690
Beech (*Fagus sylvatica* L.)	0.740

Brinell hardness tester, model HBRV-187.5E (Huatec, Beijing, China), was used. We performed the uniaxial hardness measurements in the all three main cross-sections (R, T, and L) and used two measuring force values (30 and 100 kG). Symbolic specifications of the hardness measurement conditions HB 10/294.2/60 and HB 10/980.7/60 were assigned to both sets of test conditions, respectively:

- Measuring ball diameter $D = 10$ mm
- Total force 1 $P_{30} = 30$ kG ($F = 294.2$ N, $(F/D^2_{max} = 3.2)$)
- Total force 2 $P_{100} = 100$ kG ($F = 980.7$ N, $(F/D^2_{max} = 10.6)$)
- Partial force $P_1 = 10.0$ kG (98.07 N)
- Total load time $t = 125$ s
- Number of measurements for each material $n = 12$

Figure 2 shows the Brinell tester used and the measuring force application mode.

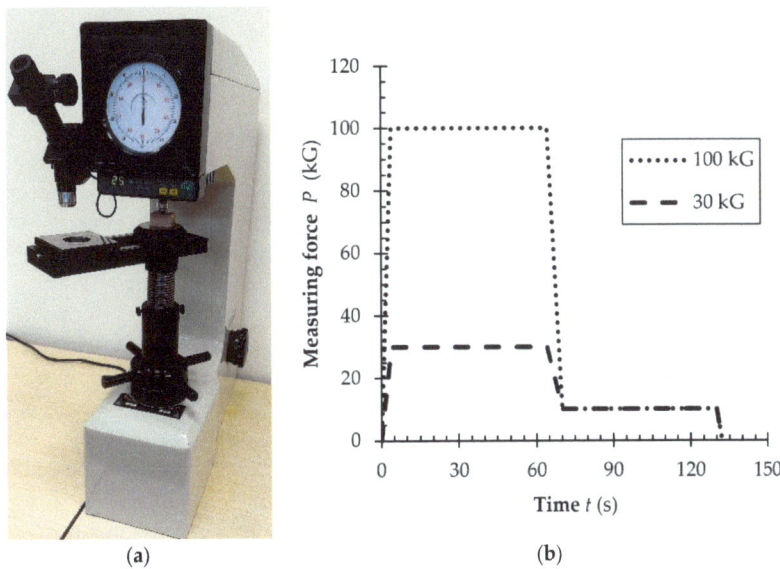

(a)   (b)

Figure 2. Hardness tester and measuring force: (**a**)—HBRV-187.5E Brinell hardness tester (Huatec, Beijing, China), (**b**)—force exertion modes.

The Brinell hardness (HB) is calculated based on the diameter of the imprint. The boundary of the imprint on the wood is unclear [8,11,12]. An additional factor that makes hardness measurement difficult is the "sinking-in effect" [12], especially in the T and R main cross-sections of wood [7]. Therefore, we used the Dino-Lite AM4815ZT EDGE digital microscope (IDCP B.V., Almere, The Netherlands) with extended dynamic range (EDR), extended depth of field (EDOF), and the possibility of measuring under polarized light. Figure 3 shows example images taken during tests.

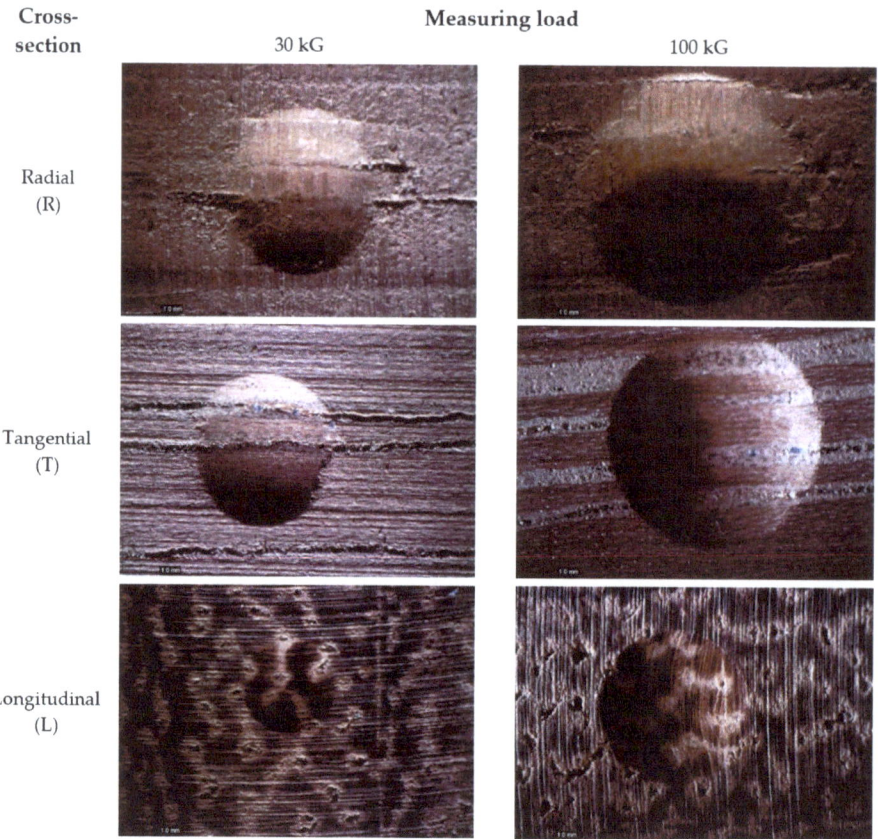

Figure 3. Examples of imprints (iroko wood, 37× magnification).

The HB values were calculated according to the following formula:

$$\mathrm{HB_d} = \frac{2 \cdot P}{\pi \cdot D \cdot \left(D - \sqrt{D^2 - d^2}\right)}$$

where:
  P = applied force (kG);
  D = diameter of the indenter (mm);
  d = diameter of the imprint (mm).

Figure 4 shows three stages of imprint creation during the hardness test: before the loading, a ball indenter under full load, and an indenter after removing the load.

The total imprint depth (H) is the sum of the depth of the permanent imprint h (the one that remains after the measuring force P is removed) and the depth of an elastic imprint x (only under load with the measuring force P). The hardness tester used allows measuring the depth of the elastic component x of the total imprint depth, which is readable after removing the measuring force ($x = H - h$) (Figure 4). Based on the indenter diameter (D), and the measured values of imprint diameter d, the permanent imprint depth can be calculated by the formula $h = D - \sqrt{D^2 - d^2}/2$. Based on the measured elastic component of the imprint (x), the total imprint depth can be calculated: $H = x + h$. Therefore, the force P and the diameter of the indenter D were constant; we measured d and x, and we calculated h and the hardness HB. Statistical calculations of the errors of the HB values and the imprint depth values were performed for the significance level of 95%: $\alpha = 0.05$, $n = 12$, 11 degrees of freedom, from the distribution of the t-Student: $t_{0.05,11} \approx 2.571$.

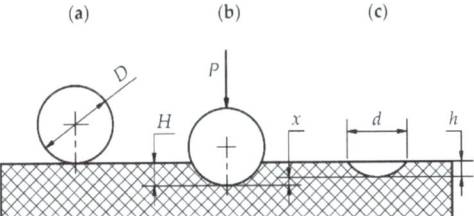

Figure 4. The imprint creation during hardness tests: (**a**)—stage 1, the ball without load, (**b**)—stage 2, the ball loaded with measuring force, (**c**)—stage 3, the permanent indentation after a force is removed (P—measuring force, D—diameter of the ball, d—the permanent imprint diameter, H—the total imprint depth, h—the permanent (plastic) imprint depth, x—the elastic (temporary) component of imprint depth).

3. Results

Figure 5 summarizes the calculated Brinell hardness values (HB) based on the diameter of the imprint (d). The wood species in Figure 5 are arranged according to their increased density. The HB of the test wood samples varied depending on the grain direction. The highest HB values were in the L cross-section, while the smaller values were in the R and T sections. Hardness also depends on the measuring force used. The HB measured at the force of 100 kG were greater than those measured at 30 kG.

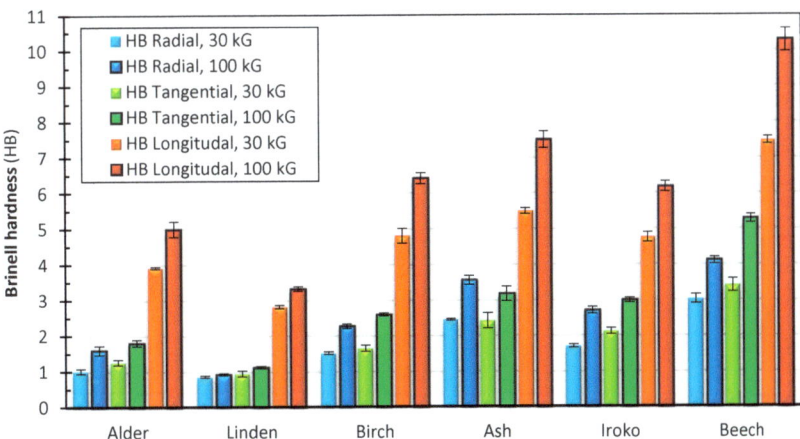

Figure 5. The Brinell hardness of the tested wood in the three main cross-sections of wood at two measuring forces (30 kG and 100 kG).

The computed hardness confidence intervals (shown as error bars in Figure 5) had varying widths. With a measuring force of 30 kG in the R cross-section, they ranged from 2% to 13% (average 7%) of hardness, in the T cross-section from 8% to 18% (average 13%), and in the L section from 1% to 9 (average 4%). However, with a measuring force of 100 kG, the confidence intervals ranged: in the R cross-section from 3% to 13% (average 12%), in the T cross-section from 5% to 15% (average 8%) and in the L cross-section from 4% to 9% (6% on average). It can be observed that the average measurement uncertainty seems to be smaller in the L cross-section than in the R and T cross-sections. The highest measurement uncertainty was calculated for linden and ash in the R section; they were 18%. The width of the confidence interval is related to the confidence level, the sample size, and the variability in the sample. We used a 95% confidence level and we performed twelve measurements for each tested combination: two measuring forces, three cross-sections, and six types of wood. The confidence intervals varied from 1% to 18%; this confirms the well-known high variability of wood properties [9].

Increasing hardness with increasing wood density was noticeable in all three cross-sections and at both measuring forces (Figure 6).

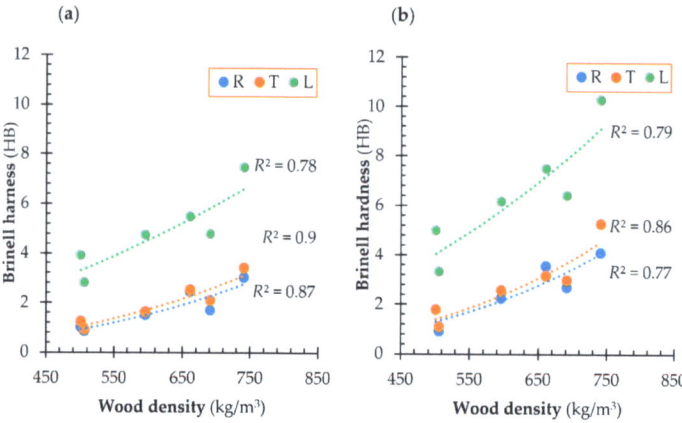

Figure 6. The Brinell hardness—wood density: (**a**)—30 kG measuring force, (**b**)—100 kG measuring force.

Figures 7–9 present the measured imprint depths. The tested wood species are arranged in ascending order according to their density, and the graphs show the two components of the total imprint depth. The permanent (plastic) imprint (h), which remains after the measuring force, is marked in blue, and the elastic component of the imprint's depth (x) is marked in green, that is, the distance by which the imprint's depth was decreased.

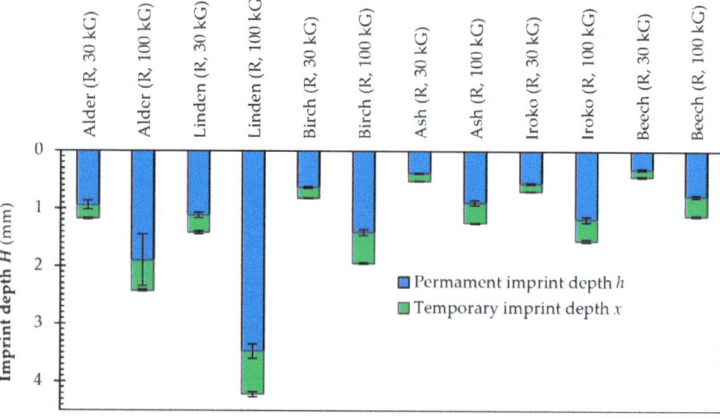

Figure 7. Imprint depth under measuring load (radial direction).

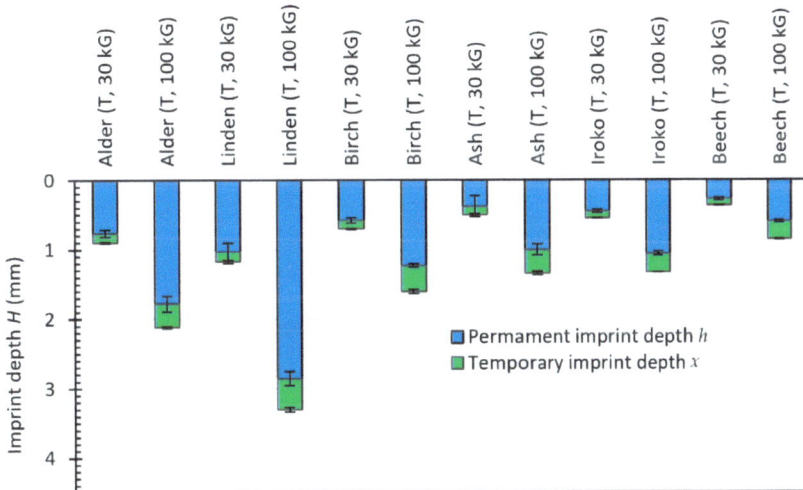

Figure 8. Imprint depth under measuring load (tangential direction).

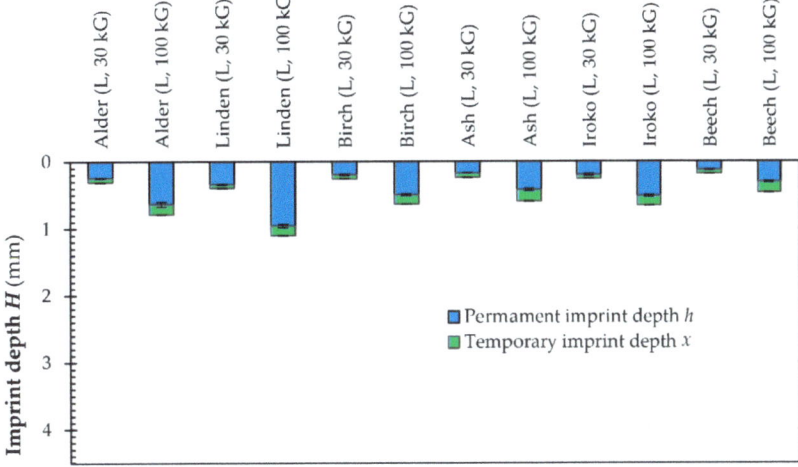

Figure 9. Imprint depth under measuring load (longitudinal direction).

Regarding the study's primary aim, the most important is the permanent imprint depth (h) shares of the total imprint depth (H). These shares are shown in Figures 10–12 (the symbols R, T, L, and h, H are set out in Figures 1 and 4).

The imprint depths caused by the force of 30 kG (Figures 7–9) were from 2.1 to 3.0-times smaller than the imprint depth caused by the 100 kG measuring force. This proportion between the imprint depths was similar in all three main wood cross-sections (R, T, L). The depths of imprints in the R and T cross-sections were similar, while they were three-times greater than in the L cross-section.

The shares of permanent imprint depth h in the total imprint depth H (including the elastic component of deformation x) did not depend on the cross-section of wood (R, T, and L) and the measuring force value. In each studied case, they ranged from 70 to 80%.

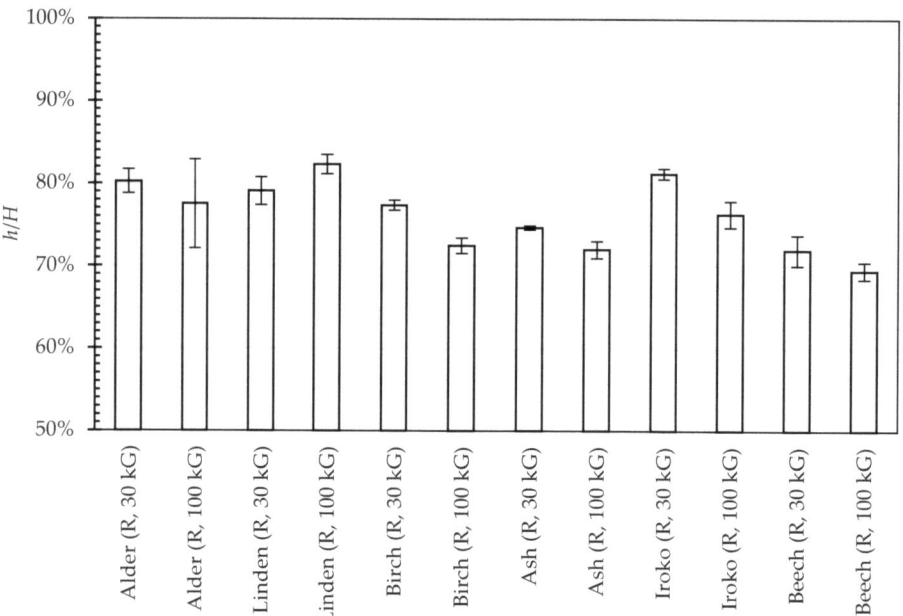

Figure 10. The R cross-section: the permanent imprint depth (h) shares the total imprint depth (H).

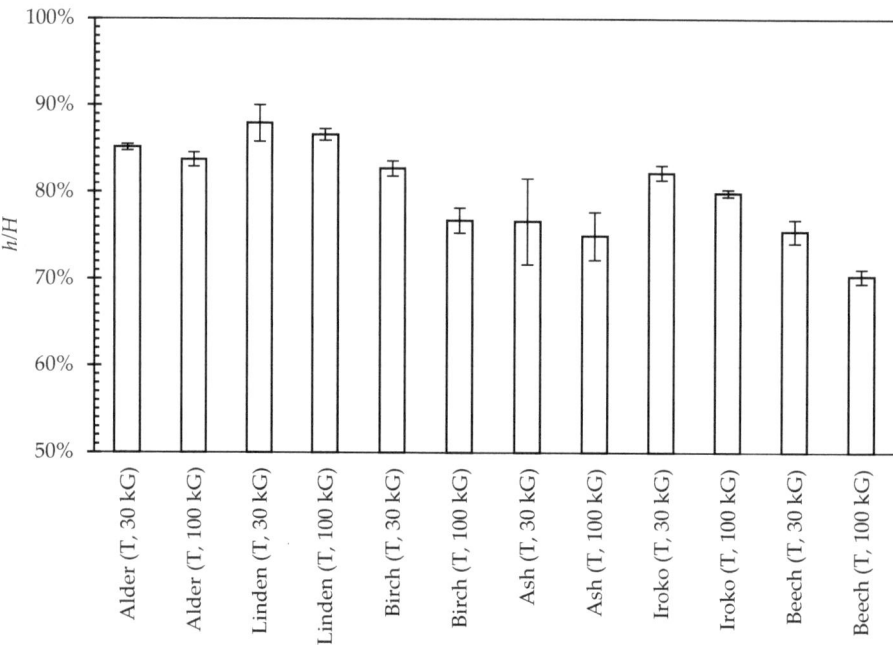

Figure 11. The T cross-section: the permanent imprint depth (h) shares the total imprint depth (H).

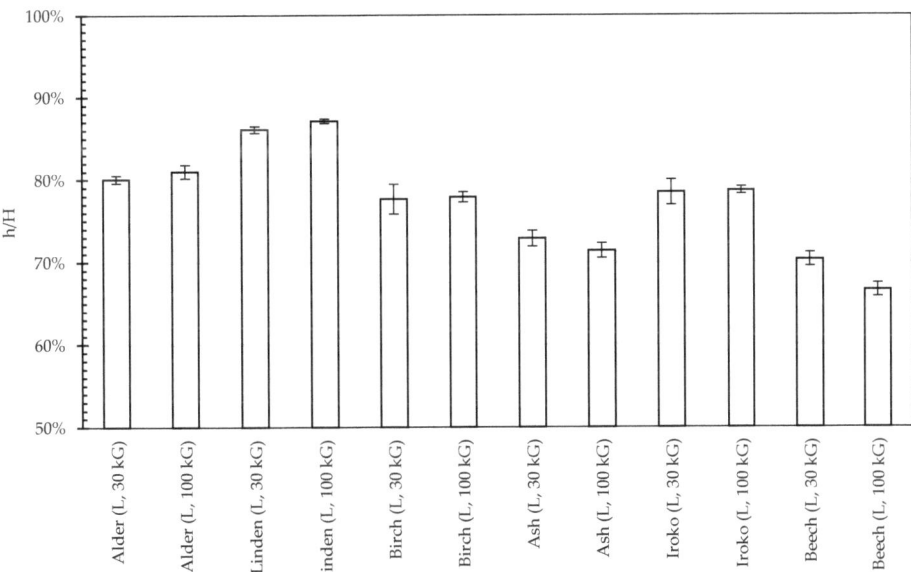

Figure 12. The L cross-section: the permanent imprint depth (h) shares of the total imprint depth (*H*).

4. Discussion

4.1. Hardness

The Brinell hardness of wood is associated with wood density [10], and it can be increased by increasing the density of the wood after pressing. Laskowska confirmed this possibility in tests involving the pressing of beech, oak, and pine wood, during which, after increasing the density of the samples by approximately 30%, the hardness increased by about 100% [13]. The hardness can also be reduced by reducing the strength of the wood after thermal modification [14]. Poplar wood, with a reduced modulus of elasticity (after thermo-modification at a temperature of 190–210 °C), had reduced hardness by about 25% [15]. Thermal modification influences the reduction of hardness in every main cross-section of wood differently: in the longitudinal cross-section by 3%, the radial cross-section by 15%, and the tangential cross-section by 25% [16]. The hardness is a property of the surface layer of the material [17], so the hardness of wood strongly depends on its density profile. Surface-densified pinewood shows an increased hardness and a high variation in measured hardness values, regardless of which testing method was used [18].

Our research confirms a well-known feature of wood mentioned in the Introduction: the hardness strongly depends on the cross-section of wood [9,10]. Our study obtained the highest hardness values in the longitudinal cross-section (L), perpendicular to the trunk axis (the end hardness of wood). The hardness was smaller in radial and tangential cross-sections (R and T—the side hardness of wood) (Figure 5). Those results are in line with the literature. For example, the radial hardness of Amboyna wood is 30–40% of the end hardness, and its tangential hardness is approximately 120–130% of the radial hardness (hardness R and T—a force of 20 kG was used; hardness L—a force of 50 kG was used) [19]. Our research confirms this regularity; hardness measured with 30 kG of all tested wood species was 26–45% and 104–124%, respectively, while hardness measured with 100 kG: 28–47% and 89–129%, respectively. In the case of beech wood, these values were 40 and 129% (force 100 kG). Similar hardness ratios were obtained in the experiment by Sedlar et al.: 42% and 126%, respectively (force 1000 N, 10 mm ball intender) [16]. Table 2 shows the ratios of hardness measured in our tests in dependence of measuring force used and in dependence of directions to wood fibers.

Table 2. HB values ratios. Three main cross-sections of wood (R, T, and L) and two measuring forces.

Ratio	Equation	Wood Specie					
		Alder	Ash	Beech	Birch	Iroko	Linden
Hardness depending on the measuring force value	$HB_{(L.\ 100\ kG)}/HB_{(L.\ 30\ kG)}$	128%	136%	138%	134%	130%	118%
	$HB_{(R.\ 100\ kG)}/HB_{(R.\ 30\ kG)}$	157%	145%	136%	150%	159%	107%
	$HB_{(T.\ 100\ kG)}/HB_{(T.\ 30\ kG)}$	143%	125%	155%	157%	142%	119%
Hardness depending on the cross-section	$HB_{(R.\ 30\ kG)}/HB_{(L.\ 30\ kG)}$	26%	45%	41%	31%	36%	30%
	$HB_{(R.\ 100\ kG)}/HB_{(L.\ 100\ kG)}$	32%	47%	40%	35%	44%	28%
	$HB_{(R.\ 30\ kG)}/HB_{(T.\ 30\ kG)}$	123%	104%	113%	109%	124%	109%
	$HB_{(R.\ 100\ kG)}/HB_{(T.\ 100\ kG)}$	104%	89%	129%	114%	111%	122%

The measured HB values also depend on the measuring force used. We obtained the higher hardness values at a force of 100 kG. Similar observations were made by Koczan et al. [20], who described the results of wood hardness tests, among others, of beech wood. A potential explanation for the higher hardness values at 100 kG than at 30 kG is the strain hardening effect, which increases with a decreasing indentation of the measuring ball. This phenomenon was observed when measuring the hardness of metals [21]; in wood, the material's cellular structure additionally influences it. Only after increasing the load did the plastic buckling of the cell walls reduce the volume of the voids and densify cell walls [22]. Based on the results of our research, the tested species can be classified into hard (ash, beech), medium-hard (alder, birch, iroko), and soft wood (linden) species. The influence of the measuring force value on the measured hardness was the lowest for soft species; it was (depending on the grain direction) from 118 to 107%. That influence ranged from 125 to 176% for the remaining wood species, as shown in Table 2. In the case of hardness measurements in the L cross-section, the influence of the force on hardness was the least diversified (118–138%); while in the R cross-section, this influence was the most diverse (107–176%).

4.2. Self-Re-Deformation

The ability to self-shallow the imprint after removing the measuring force (self-re-deformation) seems to depend on the density of the wood. A graphical representation of the self-re-deformation ability in the three main cross-sections of wood is presented in Figures 13–15.

The wood species with the highest density exhibited the highest ability of self-re-deformation. This is in line with our previous research [8] and reports from the literature [23]. In the case of the clear sapwood of kiln-dried Scots pine, the ability of side elastic self-re-deformation ranged from 45% (sphere-shaped intender, 1000 N) for densified material to 91% (cylinder-shaped intender, 2500 N) [24]. As shown in Figures 13–15, the ability to self-decrease in the depth of the imprint after removing the load in all main cross-sections (R, T and L) slightly increased with increasing wood densities. This tendency was observed for both measuring forces, 30 kG and 100 kG. In the R and T cross-sections, after the load is removed, the self-re-deformation ability was greater for the measuring force of 100 kG and less for the measuring force of 30 kG. In the L cross-section, the ability of self-re-deformation in the tested range depends only on the density of the wood (it does not depend on the value of the measuring force). These results show that the ability of self-re-deformation depends both on measuring force and the wood density in the R and T cross-sections; however, in the L cross-section, the ability of self-re-deformation depends on the wood density only. Overall, these results suggest the different progressions of cell-structure deformation in the R and T cross-sections compared to the L cross-section [25]. Cells are strongly elongated in the L direction; during compression in the R or T directions, they occur in the following sequence: (a) the linear-elastic bending of the cell walls, (b) the plastic buckling of the cell walls and reduction of void volume, and (d) cell walls are visco-elastically compressed (densification on a macroscopic level). The linear-elastic bending of the cell walls is almost fully reversible, and the densification of cells is partially reversible. When the indenter is pressed in the R and T directions, the wood cells bend and collapse after reaching their plastic collapse load. Compression in the L direction (axially) causes the kinking of elongated cell walls in the L direction [26]. Kink (failed yield) occurs by local plastic buckling [27] or by the fracture of the cells' ends [28]. Local plastic buckling usually begins at points where the cells bend to make space for a ray [25]. Vural and Ravichandran described a similar deformation process of balsa wood cells under longitudinal compression. They related the course of deformation to wood density, stating that it is by the initial elastic and then plastic buckling of cell walls in low-density specimens, while kink band formation and end-cap collapse dominate in higher-density specimens [29]. This was also confirmed by the results of our research presented in Figures 13–15. Within the wood species, the tendency to self-re-deformation was generally higher for the measuring force of 100 kG (lower h/H). The only exception was found for the softest hardwood specie (linden), where a greater tendency to self-re-deformation was observed at the measuring force of 30 kG.

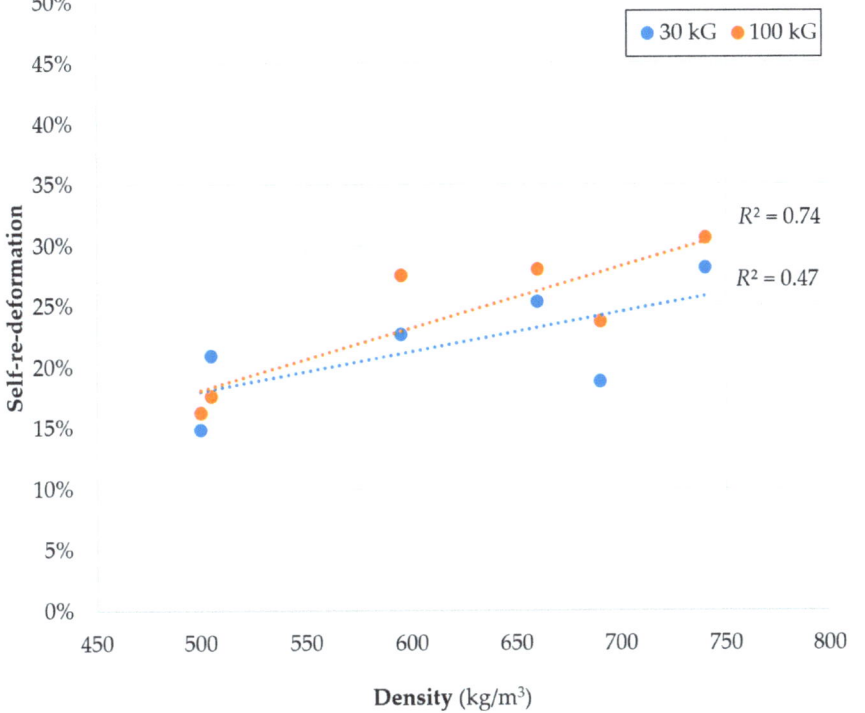

Figure 13. The ability of self-re-deformation in the R cross-section versus wood density.

The longitudinal strength of wood is always larger than the other two "directional" strengths, in part because the microfibrils of cellulose in the cell walls lie most nearly along the longitudinal direction, making the cells stiffer against longitudinal deformation [28,30]. In addition, a hexagonal prismatic wood cell is stiffer longitudinally (during compression) and less stiff transversely (in radial and tangential directions) because the thin cell walls bend [31]. The higher the density, the thicker the cell walls [9]. Therefore, density is an essential factor in predicting the strength of the wood.

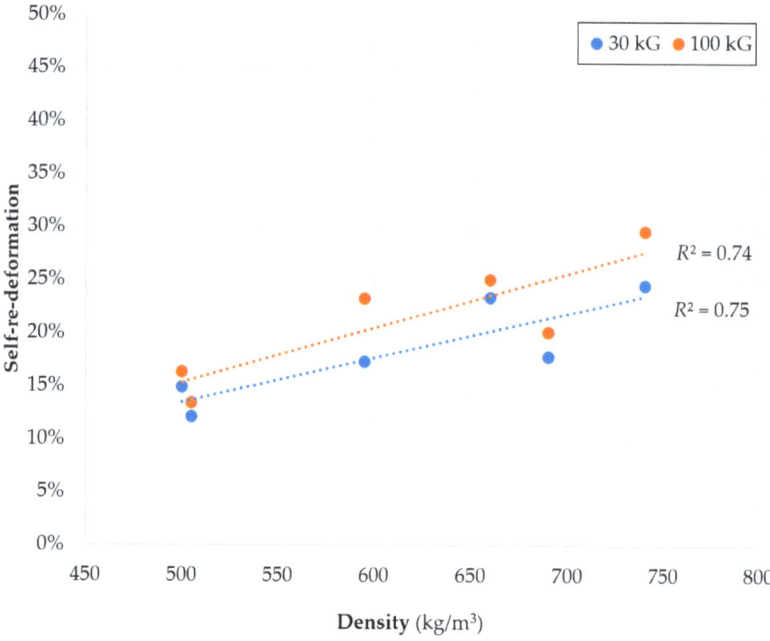

Figure 14. The ability of self-re-deformation in the T cross-section versus wood density.

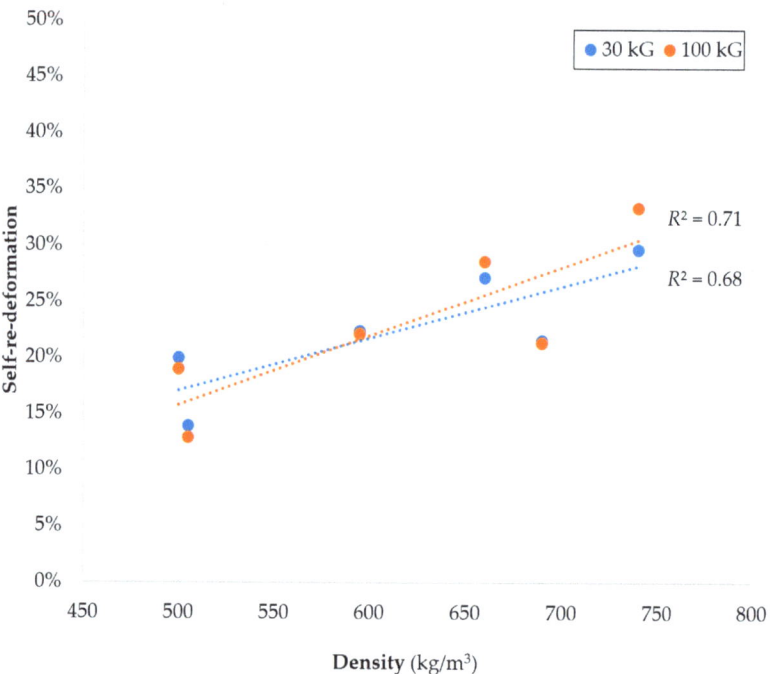

Figure 15. The ability of self-re-deformation in the L cross-section versus wood density.

5. Conclusions

- The greater the density of the wood, the highest Brinell hardness—and, at the same time, its tendency to self-re-deformation (self-shallowing of an imprint after the removal of the measuring force) is greater. The hardest tested wood species (beech) shows a share of the permanent imprint depth in the total imprint depth (h/H) only from 67 to 75%, while the wood with the lowest hardness (linden) from 82% to 88%. The self-re-deformation ability is thus linked to the wood density: the harder the wood, the smaller the share of the permanent imprint depth in the total imprint depth.
- The ability to self-re-deformation of all the tested wood species' radial and tangential cross-sections (R and T) depends on the wood density and the measuring force used. In contrast, in the longitudinal cross-section (L), this ability only depends on the wood density (the self-re-deformation ability is independent of the measuring force used). This observation shows that the compaction of the cell structure during side compression is largely reversible (semi-destructive), while the longitudinal deformation of the cell structure (the buckling of cell walls and fracture of ends of the cells) is irreversible (destructive).

Author Contributions: Conceptualization, M.S.; methodology, M.S. and G.P.; software, G.P.; validation, G.P. and M.S.; formal analysis, P.A., R.K., M.K. and T.R.; investigation, G.P.; resources, G.P. and M.S.; data curation, M.S.; writing—original draft preparation, M.S.; writing—review and editing, M.S., T.R. and P.A.; visualization, M.S. and G.P.; supervision, M.S.; project administration, P.A., R.K. and M.K.; funding acquisition, R.K. and M.K. All authors have read and agreed to the published version of the manuscript.

Funding: This research was supported by the Cultural and Educational Grant Agency of the Ministry of Education, Science, Research and Sport of the Slovak Republic under contract no. KEGA 026UMB-4/2021 and by the grant agency VEGA under project No. 1/0324/21 and project No. 1/0629/20. The study was also supported by the funding for statutory R&D activities as the research task No. 506.227.02.00 and No. 506.221.02.00 of the Faculty of Forestry and Wood Technology, Poznań University of Life Sciences.

Institutional Review Board Statement: Not applicable.

Informed Consent Statement: Not applicable.

Data Availability Statement: All data generated or analyzed during this study are included in this published article.

Conflicts of Interest: The authors declare no conflict of interest. The funders had no role in the study's design, in the collection, analyses, or interpretation of data, in the writing of the manuscript, or in the decision to publish the results.

References

1. Meyer, L.; Brischke, C.; Welzbacher, C.R. Dynamic and static hardness of wood: Method development and comparative studies. *Int. Wood Prod. J.* **2011**, *2*, 5–11. [CrossRef]
2. Liu, M.; Lin, J.; Lu, C.; Tieu, K.; Zhou, K.; Koseki, T. Progress in Indentation Study of Materials via Both Experimental and Numerical Methods. *Crystals* **2017**, *7*, 258. [CrossRef]
3. Broitman, E. Indentation Hardness Measurements at Macro-, Micro-, and Nanoscale: A Critical Overview. *Tribol. Lett.* **2017**, *65*, 23. [CrossRef]
4. Vörös, Á.; Németh, R. The History of Wood Hardness Tests. *IOP Conf. Ser. Earth Environ. Sci.* **2020**, *505*, 012020. [CrossRef]
5. EN 1534: *Wood Flooring and Parquet—Determination of Resistance to Indentation—Test Method*; European Committee for Standardization: Brussels, Belgium, 2020.
6. Niemz, P.; Stübi, T. Investigations of hardness measurements on wood based materials using a new universal measurement system. In *Symposium on Wood Machining, Properties of Wood and Wood Composites Related to Wood Machining*; Stanzl-Tschegg, S.E., Reiterer, A., Eds.; Christian-Doppler-Laboratory for Fundamentals of Wood Machining Institute of Meteorology and Physics, University of Agricultural Sciences: Vienna, Austria, 2000; pp. 51–61.
7. Lykidis, C.; Nikolakakos, M.; Sakellariou, E.; Birbilis, D. Assessment of a modification to the Brinell method for determining solid wood hardness. *Mater. Struct.* **2016**, *49*, 961–967. [CrossRef]

8. Sydor, M.; Pinkowski, G.; Jasińska, A. The Brinell Method for Determining Hardness of Wood Flooring Materials. *Forests* **2020**, *11*, 878. [CrossRef]
9. Kollmann, F.F.; Côté, W.A. *Principles of Wood Science and Technology. I Solid Wood*; Springer: Berlin/Heidelberg, Germany, 1968; ISBN 978-3-642-87930-2.
10. Mörath, E. Studien Über die Hygroskopischen Eigenschaften und die Härte der Hölzer. Habilitation Dissertation, Mitteilungen der Holzforschungsstelle an der Technischen Hochschule Darmstadt, Darmstadt, Germany, 1932.
11. Doyle, J.; Walker, J.C.F. Indentation hardness of wood. *Wood Fiber Sci.* **1985**, *17*, 369–376.
12. Hill, R.; Storåkers, B.; Zdunek, A. A theoretical study of the Brinell hardness test. *Proc. R. Soc. Lond. Math. Phys. Sci.* **1989**, *423*, 301–330.
13. Laskowska, A. Density profile and hardness of thermo-mechanically modified beech, oak and pine wood. *Drew. Prace Nauk. Doniesienia Komun.* **2020**, *63*, 26. [CrossRef]
14. Salca, E.-A.; Hiziroglu, S. Evaluation of hardness and surface quality of different wood species as function of heat treatment. *Mater. Des.* **2014**, *62*, 416–423. [CrossRef]
15. Zanuttini, R.; Negro, F.; Cremonini, C. Hardness and contact angle of thermo-treated poplar plywood for bio-building. *IForest Biogeosci. For.* **2021**, *14*, 274–277. [CrossRef]
16. Sedlar, T.; Šefc, B.; Stojnić, S.; Jarc, A.; Perić, I.; Sinković, T. Hardness of thermally modified beech wood and hornbeam wood. *Šumar. List* **2019**, *143*, 433. [CrossRef]
17. Sydor, M.; Wieloch, G. Construction Properties of Wood Taken into Consideration in Engenering Practice | Właściwości Konstrukcyjne Drewna Uwzględniane w Praktyce Inżynierskiej. *Drewno* **2009**, *52*, 63–73.
18. Scharf, A.; Neyses, B.; Sandberg, D. Hardness of surface-densified wood. Part 1: Material or product property? *Holzforschung* **2022**. [CrossRef]
19. Mijajima, H. Relation of the Hardness Number to the Applied Load in the Static Ball Indentation Test of Wood. *Nippon Rin Gakkai-Shi J. Jpn. For. Soc.* **1935**, *17*, 794–801. [CrossRef]
20. Koczan, G.; Karwat, Z.; Kozakiewicz, P. An attempt to unify the Brinell, Janka and Monnin hardness of wood on the basis of Meyer law. *J. Wood Sci.* **2021**, *67*, 7. [CrossRef]
21. Tabor, D. *The Hardness of Metals*; Oxford Classic Texts in the Physical Sciences; Clarendon Press, Oxford University Press: Oxford, UK; New York, NY, USA, 2000; ISBN 978-0-19-850776-5.
22. Gibson, L.J.; Ashby, M.F. *Cellular Solids: Structure and Properties*, 2nd ed.; Cambridge University Press: Cambridge, UK, 1997; ISBN 978-0-521-49911-8.
23. Heräjärvi, H. Variation of basic density and Brinell hardness within mature Finnish Betula pendula and B. pubescens stems. *Wood Fiber Sci.* **2004**, *36*, 216–227.
24. Rautkari, L.; Laine, K.; Kutnar, A.; Medved, S.; Hughes, M. Hardness and density profile of surface densified and thermally modified Scots pine in relation to degree of densification. *J. Mater. Sci.* **2013**, *48*, 2370–2375. [CrossRef]
25. Ashby, M.F.; Jones, D.R.H. *Engineering Materials 2: An Introduction to Microstructures, Processing and Design*, 3rd ed.; Butterworth-Heinemann: Oxford, UK, 2006; ISBN 978-0-08-046863-1.
26. Adusumalli, R.-B.; Raghavan, R.; Ghisleni, R.; Zimmermann, T.; Michler, J. Deformation and failure mechanism of secondary cell wall in Spruce late wood. *Appl. Phys. A* **2010**, *100*, 447–452. [CrossRef]
27. Zauner, M.; Stampanoni, M.; Niemz, P. Failure and failure mechanisms of wood during longitudinal compression monitored by synchrotron micro-computed tomography. *Holzforschung* **2016**, *70*, 179–185. [CrossRef]
28. Gibson, L.J. The hierarchical structure and mechanics of plant materials. *J. R. Soc. Interface* **2012**, *9*, 2749–2766. [CrossRef] [PubMed]
29. Vural, M.; Ravichandran, G. Microstructural aspects and modeling of failure in naturally occurring porous composites. *Mech. Mater.* **2003**, *35*, 523–536. [CrossRef]
30. Thibaut, B.; Gril, J.; Fournier, M. Mechanics of wood and trees: Some new highlights for an old story. *Comptes Rendus Acad. Sci. Ser. IIB Mech.* **2001**, *329*, 701–716. [CrossRef]
31. Easterling, K.; Harrysson, R.; Gibson, L.; Ashby, M.F. On the mechanics of balsa and other woods. *Proc. R. Soc. Lond. Math. Phys. Sci.* **1982**, *383*, 31–41. [CrossRef]

MDPI AG
Grosspeteranlage 5
4052 Basel
Switzerland
Tel.: +41 61 683 77 34

Applied Sciences Editorial Office
E-mail: applsci@mdpi.com
www.mdpi.com/journal/applsci

Disclaimer/Publisher's Note: The statements, opinions and data contained in all publications are solely those of the individual author(s) and contributor(s) and not of MDPI and/or the editor(s). MDPI and/or the editor(s) disclaim responsibility for any injury to people or property resulting from any ideas, methods, instructions or products referred to in the content.

www.ingramcontent.com/pod-product-compliance
Lightning Source LLC
LaVergne TN
LVHW072358090526
838202LV00019B/2574